A FIRST CHEMISTRY COURSE

E N Ramsden BSc, PhD, DPhil

Wolfreton School, Hull

Stanley Thornes (Publishers) Ltd

First published in 1980 by Stanley Thornes (Publishers) Ltd, Educa House, Old Station Drive, off Leckhampton Road, Cheltenham GL53 0DN

Reprinted 1980 with minor corrections
Reprinted 1981
Reprinted 1983

British Library Cataloguing in Publication Data
Ramsden, E N
    A first chemistry course
    1. Chemistry
    I. Title
    540   QD33
    ISBN 0-85950-488-3

Typeset by Malvern Typesetting Services Ltd, Malvern
Printed in Great Britain by The Pitman Press, Bath

Contents

Preface

This book is an introduction to the study of Chemistry. It is designed to cater for the needs of a pupil who is completely new to the subject and take him or her up to the beginning of an examination course. In many schools, the book will cover a period of three years, between the ages of 11 and 14. The book is suitable for use in secondary schools, middle schools and preparatory schools. It may be used in the study of Chemistry either as a separate subject or as the chemical part of a General Science course.

For many pupils, the course followed in this book will lead on to a course of study for the CSE or GCE examinations. The section on formulae, equations and bonding in Chapter 7 is included for the sake of pupils who will go on to study Chemistry at examination level. Each chapter ends with a set of questions and a crossword..

For all pupils, Chemistry offers the opportunity to experience the enjoyment of working at chemical experiments. The sights and sounds and smells of Chemistry can be fun for pupils of all abilities. The experiments are presented in detailed form, with the steps the pupil must take clearly numbered. A discussion of results is included, but, in order to maintain a basis of discovery by the pupils, the experimental section is kept separate from the main text. Pupils can therefore work through an experiment and form their own conclusions, before checking their results against the discussion in the main text. The main text can however be read independently of the experiments, either for information or for revision.

All the experiments may be done by pupils, except those designated as Demonstration Experiments. A very large number of experiments use only the simplest of apparatus.

E. N. RAMSDEN
Hull 1980

Acknowledgements

I would like to thank the following firms which have kindly supplied photographs for inclusion in this book:

ICI Ltd. for Figure 2.1
Distillers Company for Figure 2.9
British Oxygen Company for Figures 5.12 and 5.13
Blue Circle Cement Company for Figures 6.5 and 6.12

The photograph used in Figure 8.12 is Crown copyright and is reproduced with the permission of the Controller of Her Majesty's Stationery Office.

I would like to acknowledge my debt to all those who have helped me at various stages during the preparation of this book: Dr G. N. Gilmore for reading the original draft and making corrections and suggestions which have improved the precision and clarity of the text; Dr J. Bradley for Experiment 6.5; Miss Judy Slator for typing the manuscript; Stanley Thornes (Publishers) for the close attention which they have given to every detail involved in the production of this book. Finally, I thank my family for the help and encouragement they have given me.

E. N. Ramsden
Hull, 1980

1. Working at Chemistry

1.1 What is Chemistry?

Over the centuries, man's natural curiosity has led him to find out more and more about the world, and a vast body of knowledge has been built up. This knowledge is divided into sections, such as *Physics*, *Geology*, *Astronomy* and *Chemistry*. All these sections are called *Sciences*. A science is an organised body of knowledge.

Chemistry is the study of the matter of which the world is composed. Chemists isolate pure substances from the world around us. A substance is one particular kind of matter. Chemists analyse substances to find out what they are made of and study their properties (that is, their appearance and behaviour). They also study chemical reactions in which new substances are formed and find uses for them. Plastics, detergents, soaps, antiseptics, anaesthetics, antibiotics and many other substances have all been discovered as the result of chemists' work. The work involved in discovering something new is called *research*. No one can say which lines of research will lead to most discoveries or which will lead to the most useful discoveries. Often the most important discoveries (such as X-rays and penicillin) have been made by scientists who were guided solely by the fascination of finding out more and more about their special subjects.

Discoveries have always been the result of work by scientists with trained minds. Scientific training is essentially a training in observation and deduction, rather like detective work. Scientists ask questions, do experiments designed to answer a particular question, and make observations. They record what they see and the results of the measurements they make. After observation comes deduction. Deduction means thinking about what information can be obtained from the measurements. What do their results tell them about the problem they are working on? From the information they have gained, the scientists may be able to work out a theory about how things happen. Then they will want to test their theory with fresh experiments.

Scientists do their experiments in laboratories. A *laboratory* (or *lab*) is a room equipped with all the special facilities that are needed. There are rules for *lab* work, which all scientists observe.

1.2 Working in a Chemistry laboratory

Here is a list of rules to be followed when working in a Chemistry laboratory.

Laboratory Rules

1. Concentrate on your own experiment. Do not tamper with equipment which is not part of your experiment. Do not move around more than necessary. Move at a reasonable pace.

2. Wear safety glasses. If your hair falls forward, tie it back.

3. Follow directions carefully. Make sure you are using the correct chemicals in the correct quantities. Do not do experiments of your own devising without checking with your teacher.

4. Do not taste chemicals. Smell gases cautiously.

5. When heating a chemical in a test tube, be sure you are not pointing the test tube towards yourself or another pupil.

6. Write down your observations as soon as you have made them.

7. Use clean apparatus. Wash up and tidy up after a practical lesson. Put solid waste into the bins, not into the sinks. Let hot objects cool before putting them away.

8. In case of accident, a burn, a cut or a splash of some chemical, wash with plenty of cold water. Inform your teacher immediately.

1.3 Apparatus: the Bunsen burner

Apparatus is the word given to the whole array of equipment which scientists use in their work. All the glass vessels, tools and measuring devices would be described as apparatus.

The piece of apparatus we most often use for heating is the Bunsen burner. Figure 1.1 shows the design of this gas burner, which was invented in 1854 by a German chemist called Wilhelm Bunsen.

The chimney of the Bunsen burner screws into the base plate around the pin-point jet through which gas enters. The chimney has a circular hole in it, opposite the jet. A collar which fits round the bottom of the chimney has a circular hole at the same level.

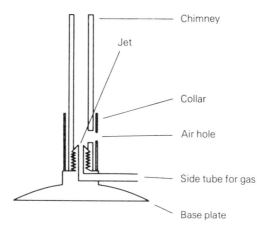

Figure 1.1   The Bunsen burner

When the hole in the collar is opposite the air hole in the chimney, air can enter. By rotating the collar, the air supply can be regulated, and different types of flame obtained.

Experiment 1.1 will give you practice in using the Bunsen burner. With the air hole closed, you get a yellow flame. This flame is luminous: it gives out light. This is because, with the air hole closed, the gas has not been fully burnt, and unburnt particles of carbon in the flame become hot and glow.

Experiment 1.1

To light a Bunsen burner

1.   Connect the Bunsen to a gas tap. Close the air hole. Open the gas half way to the fully open position, and apply a lighted splint to the top of the chimney.

2.   Note the appearance of the flame. Make a drawing.

3.   Hold a piece of porcelain, with tongs, in the flame. Observe the deposit which forms on the cold porcelain.

4.   Slowly open the air hole. Notice the change in the amount of heat and light given out, the length and temperature of the flame and the noise. Draw the flame.

5.   With the air hole open, turn the gas fully on. Note the appearance and the sound of the flame.

6.   Compare your drawings with those in Figure 1.2.

Experiment 1.2

To find out the action of heat on some substances

1. Quarter fill an ignition tube with one of the substances listed below in step 4. For iodine, however, use only one flake. Hold the ignition tube with tongs. Point it away from your neighbour and yourself.

2. Heat the ignition tube in a blue Bunsen flame. Observe carefully.

3. Write down your observations in the form of a table.

Table 1.1 *Action of heat on some substances*

Substance heated		Observations
Name	Appearance	

4. One at a time, heat the remainder of the following substances: wax, ice, water, zinc oxide, iodine (one flake), ammonium chloride, copper carbonate, copper sulphate crystals.

5. Look at your table of results, and ask yourself which of the changes that occur on heating are easily reversed, to get back to the starting material, and which are difficult to reverse.

When you put a piece of cold porcelain into the flame, the particles of carbon are deposited on the cold surface as soot. Whenever you want to leave a Bunsen burner lit and unattended, close the air hole so that you get the yellow flame, which is clearly seen, and which no one will stick an elbow into by mistake.

When you want to heat something, as you do not want a deposit of soot, open the air hole to obtain the blue flame. A 'half and half' flame, with the gas half way on and the air hole half open, gives a general purpose flame. Use this flame unless your instructions ask for strong heating, then you can fully open the air hole and turn up the gas to get a roaring flame. The blue flame has an inner zone of unburnt gas, a blue zone of incomplete combustion, and a pale blue outer cone of complete combustion. The hottest part of the flame is just above the top of the blue zone.

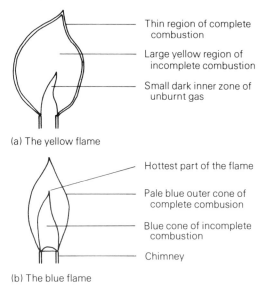

(a) The yellow flame

— Thin region of complete combustion

— Large yellow region of incomplete combustion

— Small dark inner zone of unburnt gas

(b) The blue flame

— Hottest part of the flame

— Pale blue outer cone of complete combusion

— Blue cone of incomplete combustion

— Chimney

Figure 1.2   The Bunsen burner flames

Remember, when using a Bunsen burner:
(1) *to light*, have the air hole closed and the gas half on;
(2) *to leave unattended*, use the yellow flame, with the air hole closed;
(3) *to heat*, use the blue flame, with the air hole half open and the gas half on;
(4) *to heat strongly*, use the roaring flame, with the air hole wide open and the gas fully on.

The Bunsen burner was designed to burn coal gas. North Sea gas is now a more popular fuel. It gives out more heat than coal gas, and a Bunsen burner with a finer gas jet and a thicker chimney is used.

You can now use a Bunsen burner to heat some chemicals, as described in Experiment 1.2.

5

1.4 The action of heat on some substances; physical and chemical changes

Experiment 1.2 is to find out what happens when you heat various substances. In some cases, there is a temporary change when the substance is heated, but, on cooling, the original substance is formed again. In other cases, there is a permanent change on heating, and a new substance is formed.

You will have found out that the solid wax simply melts on heating to liquid wax, and the change is reversed by cooling. The same is true of ice, which melts to form water on heating, and can be obtained from water by cooling it until it freezes. We can describe these changes by writing

Melting

Heat

Solid $\rightleftharpoons$ Liquid

Cool

Solidification

where each arrow stands for 'forms'.

Water turns into steam on heating. In general, a liquid turns into a gas on heating. This process is called evaporation because a gas at a temperature close to that at which it becomes a liquid is called a *vapour*. The reverse of this process is changing a vapour into a liquid. This is called *condensation*.

Evaporation

Heat

Liquid $\rightleftharpoons$ Vapour

Cool

Condensation

Changes like melting and evaporation, which are changes from a substance in one form to the same substance in a different form, are called *physical changes*: no new substances are formed.

When zinc oxide is heated, you observe another physical change which is easily reversed by cooling: it is a change from white zinc oxide to yellow zinc oxide.

When iodine is heated, you see the shiny black crystals change to a purple vapour, and at the cooler end of the ignition tube you see black crystals form again. The whole process from solid to vapour on heating and from vapour to solid on cooling is called *sublimation*.

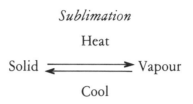

Sublimation

Heat

Solid ⇌ Vapour

Cool

It is unusual. Most solids melt on heating, and then change to a vapour on strong heating. In sublimation, the expected liquid stage is missed out. You see sublimation again when ammonium chloride is heated. It is a physical change.

When you heat blue crystals of copper sulphate, you may notice steam coming out of the ignition tube; then a white solid is left in the tube. It looks as though a chemical reaction has occurred, to give two substances which are different from the one you started with. Let the tube cool, and add a few drops of cold water. You will see the blue colour return and feel the tube become hot. The reaction:

heat + blue copper sulphate ⇌ steam + white copper sulphate
       crystals   powder

will go from left to right or from right to left. It is a *chemical change* because new substances are formed. It is easily reversed.

When you heat copper carbonate, you see the green solid change to a black solid. The product looks quite different from the starting material, and in fact it is chemically different. For example, it does not fizz when put into a dilute acid as does copper carbonate. We conclude that a chemical change has taken place. This change is not easily reversed, but by a series of chemical reactions you can get back to copper carbonate.

Ask your teacher to heat some ammonium dichromate in the fume cupboard. The orange crystals change spectacularly to a solid which looks quite different. The reaction is so vigorous that you will be sure you have been watching a chemical change occurring.

These tests will have given you an idea of the difference between a physical change and a chemical change. In a *physical change*, the form of a substance changes, but it is still the same substance. In a *chemical change*, new substances are formed. The physical changes met so far are *changes of state*. Matter exists in three states, solid, liquid and gas.

1.5 States of matter

Solid
A solid has a definite size and a definite shape. Solids may increase slightly in size when heated: you may know of the expansion of metal rods on heating. The shape may be changed when force is applied: for example, the bumper of a car may be dented if it runs into a concrete post.

Liquid
A liquid has a definite size, but no definite shape: it adopts the shape of its container. If you take a measuring cylinder containing 50 cm³ of liquid, the liquid will have a certain shape. Tip it into a 50 cm³ beaker, and it will have a different shape; tip it into a 250 cm³ beaker, and the liquid will have a third shape, as Figure 1.3 shows. Liquids expand on heating more than solids do.

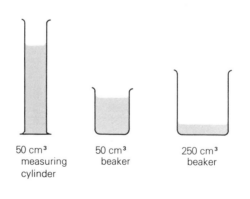

50 cm³
measuring
cylinder

50 cm³
beaker

250 cm³
beaker

Figure 1.3   Three shapes for 50 cm³ of liquid

Gas   A gas has neither a definite size nor a definite shape: it has the size and the shape of its container. Gases always spread out to occupy all the space available to them. If you take a balloon full of gas and 'pop' it, the gas inside it will spread out evenly all over the room. This behaviour of gases is called diffusion. Gases expand a great deal on heating if the container allows expansion. You can take a dent out of a table tennis ball by warming it gently so that the air inside expands and pushes the dent out. If the container is rigid, the gas cannot increase in volume, and the pressure of gas inside the container increases.

Water is a substance which we know in all three states, as a solid, a liquid and a gas. We call these *ice*, *water* and *steam*. Water is the same chemical in all three states, and its chemical reactions are the same. How does it come about that a substance can exist in three forms which are physically different? You will be able to answer this question after we have studied the idea that matter is made up of tiny particles called *atoms* (see Chapter 7).

Questions on Chapter 1

1. Write down the correct words to fill in the blanks in these sentences.

   (a) Ammonium chloride changes from solid to gas on heating. This process is called _evaporation_

   (b) Copper carbonate is _green_ in colour. It becomes _black_ on heating.

   (c) When zinc oxide is heated, it becomes _yellow_ and then returns to white on cooling.

   (d) A black, crystalline solid forms a purple vapour when it is heated. This solid is _iodine_.

   (e) Copper sulphate crystals are _blue_ in colour. When they are heated they turn _white_ and give off water. If water is added to the solid left behind, the colour changes to _blue_.

   (f) A solid has a definite _size_ and a definite _shape_.

9

2. (a) What form of matter has a definite volume but no definite shape? *liquid*

(b) Explain what is meant by the terms *evaporation*, *condensation* and *sublimation*.

3. Explain what is meant by a physical change and what is meant by a chemical change. Give two examples of each type of change.

4. What are the three states of matter? Name one substance which can exist in all three states, and say how it can be changed from one state to another.

Crossword on Chapter 1

Across

1   1 down. Used for heating in the laboratory (6, 6)
5   Try to ____ accidents in the laboratory! (5)
7   A state of matter (5)
9   Pleasant (4)
11 Place for animals in a biology laboratory? (4)
12 An attractive alloy containing copper (6)
14 A form of movement not allowed in the laboratory! (7)
15, 19 across. Needed to make a new substance (8, 6)
16 Find the position of (6)
17 Develop these for a strong body (7)
18 The smallest (5)
19 See 15 across and 13 down
20 Another state of matter (6)
21 Used for weighing (7)

Down

1   See 1 across
2   Behave in a ____ (not dangerous) way in the laboratory (4)
3   Use these for holding hot objects (5)
4   Put rubbish in here (3)
6   Change from liquid to gas (8)
8   Leave the flame from 1 across, 1 down this colour for visibility (6)
10 This flask sounds almost amusing! (7)
11 Changed from vapour to liquid (9)
13, 19 across. When this happens, no new substance is formed (8, 6)
16 Allowed by law (5)

Trace this grid on to a piece of paper, and then fill in the answers.

2. Methods of separating mixtures

This chapter deals with the various methods that can be used to separate a mixture of two or more substances into its different **parts.** Chemists often need to do this when they want to obtain pure **substa**nces from the world around us.

2.1 Filtration

A mixture of a solid and a liquid is easy to separate if you have a sieve which will allow the liquid to pass through and retain the solid particles. Experiment 2.1 is an example of this method. It uses the process called filtration. *Filtration is the process of separating solid particles from a liquid by allowing the liquid to pass through some porous material.* Material is said to be porous when it acts like a very fine sieve, allowing liquid to pass through but retaining solid particles. The liquid which passes through is called the *filtrate*, and the solid which remains in the filter is called the *residue*. We often use a porous type of paper called *filter paper*, and support it in a glass funnel called a *filter funnel*.

You may have the problem of separating two solids, of which only one dissolves in water. You can tackle this problem by first dissolving one of the solids and then filtering to separate the solution from the insoluble solid. Sand and salt form just such a mixture, and Experiment 2.2 shows you how to separate them.

Rock salt is mined in Northwich, Cheshire, and Figure 2.1 shows a truck laden with about 25 tons of rock salt, driving through the mine. The salt is taken to the ICI factory at Winnington where pure salt is extracted from the rock salt. The pure salt is used in the manufacture of washing soda and other chemicals.

Experiment 2.1

To separate soil and water

1. Take a mixture of soil and water.

2. Prepare a filter funnel and filter paper in a clamp and stand. Put a beaker underneath the filter funnel, as in Figure 2.2.

3. Pour the mixture into the funnel, making sure that it goes into the filter paper and not straight down the funnel.

4. Inspect the filtrate, and see whether you have succeeded in separating the soil from the water.

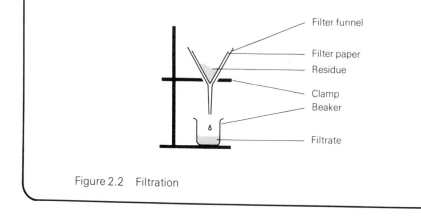

Figure 2.2   Filtration

Figure 2.1   A salt mine in Cheshire

To obtain pure salt from rock salt

1. Crush and grind the rock salt in a mortar with the help of a pestle (see Figure 2.3 (a)).

2. Place the solid in a beaker. Half fill the beaker with water.

3. Stand the beaker on a tripod and gauze. Heat gently (gas tap half on, air hole half open). Stir with a glass rod to help the salt to dissolve (see Figure 2.3 (b)).

4. Pour the contents of the beaker through a filter funnel fitted with a filter paper (see Figure 2.3 (c)). This will remove undissolved impurities such as sand. Collect the filtrate in an evaporating dish.

5. Place the dish on a steam bath and evaporate to dryness, as in Figure 2.3 (d).

6. In this experiment, as an exception to our safety rules, you may taste the product.

   Do you think that the white crystalline solid in the evaporating basin is salt?

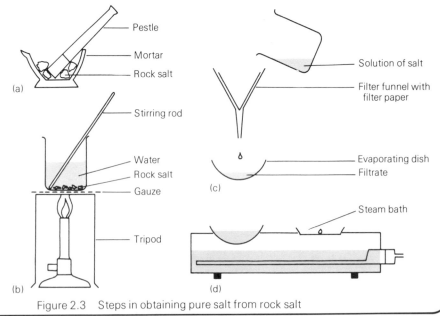

Figure 2.3   Steps in obtaining pure salt from rock salt

Some new words have been met in this experiment. When we say that a solid *dissolves* in a liquid, we mean that the solid spreads out, in minute particles, throughout the liquid so that we can no longer say which is the solid and which is the liquid. The mixture of solid and liquid is called a *solution*. The process is called *dissolving* or *solution*. The solid is the *solute*, and the liquid the *solvent*. Liquids and gases can dissolve in liquids. A substance which will dissolve is said to be *soluble*; one which will not dissolve is *insoluble*. Solution is a change of state; it is a physical change: no chemical reaction occurs. The solute can be recovered from the solution by evaporating the solvent.

Evaporation is the process of changing a liquid or a solid into a vapour. When a substance is a solid or a liquid at room temperature, we call the gaseous form of that substance a *vapour*.

Solvents other than water

Water is the most common solvent. There are many substances which do not dissolve in water and yet are required in solution. Other solvents must be used. You may have disinfected minor cuts with tincture of iodine, which is a solution of iodine in *ethanol*. A solution of shellac in ethanol is used as a clear varnish. The rubber solution that you use to repair punctures is a solution of rubber in *trichloroethene*, a solvent rather similar to *chloroform*. Nail varnish is dissolved in *pentyl ethanoate*, a sweet smelling solvent. *White spirit*, which is obtained from petroleum oil, is a solvent for paint. It is used for diluting oil paints and for dissolving the paint from brushes. You can probably think of many more solvents which you use at home and in your hobbies.

2.2 Sublimation

When trying to separate a mixture of two solids, you may find that one of them will sublime – vaporise on heating (see Chapter 1.4) – and the other will not. This difference will allow you to separate them. Experiment 2.3 shows how you can heat a mixture of solids and obtain one of them from the vapour, while the other remains in the heated vessel.

To separate a mixture of iodine and carbon

1. Assemble the apparatus shown in Figure 2.4. A piece of pressure tubing (external diameter 25 mm, internal diameter 16 mm) around the test tube will hold it in position inside a boiling tube, giving a loose fit and allowing hot air to escape. Put the mixture into the boiling tube. Fill the test tube with cold water, and clamp the boiling tube.

2. Heat gently. If the water becomes hot, stop heating, wait until the apparatus is cool enough to handle, tip out the hot water, and replace it with cold water.

3. Make a note of the colour of the iodine vapour in the boiling tube.

4. Make a note of the colour of the shiny crystals of iodine which you can scrape off the sides of the cold test tube.

5. If you have a side-arm tube, you can assemble the improved apparatus for sublimation shown in Figure 2.5. The continuous supply of cold water means that you do not have to refill the test tube. This device is called a *cold finger*.

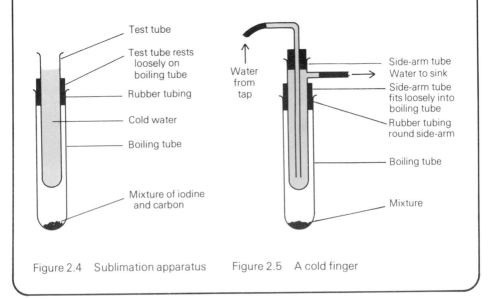

Test tube

Test tube rests loosely on boiling tube

Water from tap

Side-arm tube
Water to sink

Rubber tubing

Side-arm tube fits loosely into boiling tube

Cold water

Rubber tubing round side-arm

Boiling tube

Boiling tube

Mixture of iodine and carbon

Mixture

Figure 2.4   Sublimation apparatus

Figure 2.5   A cold finger

2.3 Separation of immiscible liquids

Two liquids which do not mix (are immiscible), such as oil and water, can be separated by a device called a separating funnel. This is shown in Figure 2.6.

The mixture of, say, oil and water is poured into the separating funnel. It is left to settle until there is a clear dividing line between the two layers. The stopper is removed. Then the tap is opened, and water is allowed to run out slowly into a beaker. When all the water has run out and the oil layer is just reaching the tap, the tap is closed. Oil is now in the funnel and water in the beaker. The narrow part of the funnel, A, below the tap, contains water. By opening the tap gently, this small amount of water can be run to waste. After substituting a clean beaker, the tap is opened again to let oil run into the second beaker.

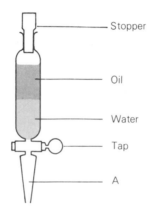

Figure 2.6   A separating funnel

2.4 Separation of miscible liquids by distillation

It is possible to obtain the solvent from a solution by heating the solution until the solvent *evaporates* (changes from liquid to vapour) and then cooling the vapour until it *condenses* (changes from vapour to liquid). *The process of evaporation in one part of the apparatus followed by condensation in another part of the apparatus is called distillation.*

In addition to being used to separate a solvent from a solute, distillation can be used to separate a mixture of two or more liquids, provided they distil over at different temperatures. Experiments 2.4 and 2.5 involve distillation.

17

To find out whether ink can be separated into components

1. First try the filtration method. Does it work?

2. Try boiling some ink in a test tube. If you see that a vapour is formed, set up an apparatus in which you can trap the vapour and condense it. Figure 2.7 shows one possible apparatus.

3. Put ink into a conical flask. Drop in some pieces of broken porcelain (such as bits of a broken evaporating basin). These will help the liquid to boil smoothly.

4. Fit the flask with a well-fitting cork and a long delivery tube, which ends inside a test tube which is cooled by standing in a beaker of cold water.

5. Heat the ink with a 'half-and-half' flame.

6. Observe the distillate collected in the test tube. Does it look like ink?

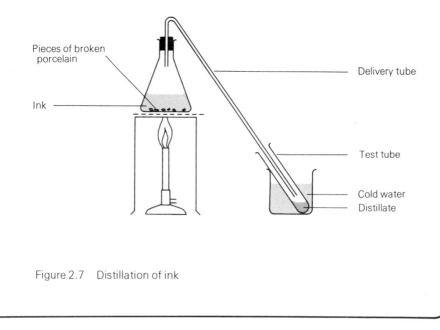

Figure 2.7   Distillation of ink

If you have tried Experiment 2.4 on ink, you will have found out that your distillate is a colourless liquid which looks more like water than ink. If you go on distilling for long enough, all the liquid will distil over and a coloured solid will remain in the conical flask.

You will have noticed that the method of cooling the vapour is not good enough, and some steam escapes. A chemist called Liebig noticed this over a century ago and designed a better method of cooling the vapour. It is called a Liebig condenser. It consists of a tube for the vapour to pass through surrounded by a wider tube, which is kept filled with cold water. The cold outer tube will condense vapour in the inner tube. Cold water constantly passes in from the cold tap at the bottom end of the outer tube and out from the top end of the outer tube into the sink. Figure 2.8 shows a Liebig condenser. You can use this in Experiment 2.5

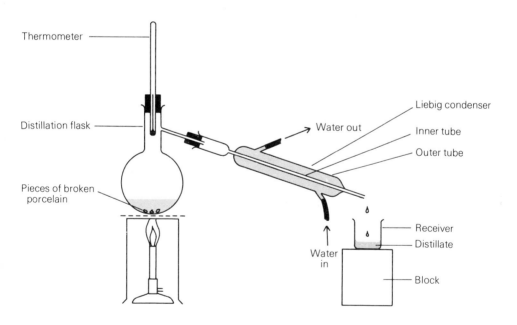

Figure 2.8   Distillation flask and Liebig condenser

To distil a mixture of ethanol and water, using a Liebig condenser

1. Put the water–ethanol mixture into the flask shown in Figure 2.8. This is called a distillation flask because of its design. The round bottom ensures even heating of the contents, and the side-arm lets out the vapour.

2. Add some pieces of broken porcelain. Clamp the flask over a tripod and gauze. Fit a thermometer and cork, with the bulb of the thermometer just below the side-arm.

3. Bring the Liebig condenser into position carefully, and clamp it at the correct height.

4. Connect pieces of rubber tubing from the inlet of the condenser to the tap and from the outlet to the sink. Turn on the tap so that a steady trickle of water flows through the condenser.

5. Have a receiver ready to collect any distillate.

6. Heat the contents of the flask. You will see the temperature recorded by the thermometer rise until it reaches 78 °C. It remains steady at this reading while a colourless liquid distils over.

7. When the temperature rises from 78 °C, you will notice that all the ethanol has distilled over. The temperature will rise to 100 °C, and remain at 100 °C while water distils over. Use a separate receiver to collect the water.

8. Do the following tests on both distillates:

    (a) Smell it.

    (b) Taste a drop on the end of your finger.

    (c) Take 2 cm³ with a teat pipette, put on to a watch glass, and test with a lighted splint.

On separating a mixture of ethanol and water by distillation as in Experiment 2.5, you will have found four differences between ethanol and water, (a) the smell, (b) the taste, (c) the *flammability* (ability to burn), (d) the difference in the boiling points.

The reason why ethanol separates from water is that ethanol has a lower boiling point (78 °C) than water (100 °C). All the ethanol in the mixture distils over before any water starts to distil.

Distillation is used in the preparation of whisky. Figure 2.9 shows the stillhouse at Ord Distillery, operated by Scottish Malt Distillers for the Distillers Company.

The method of separating two or more liquids by distillation is very important in industry. If there are a number of liquids with boiling points fairly close together, a distillation flask with a long column instead of a side-arm is used. The column provides a large surface over which evaporation and condensation can take place many times. This improves the efficiency of separating the mixture into its parts or fractions. Such a column is called a *fractionating column*. One of the most important users of fractional distillation is the oil industry. It separates crude petroleum oil into many fractions with different boiling points by this method (see Chapter 6.8).

Figure 2.9   In the stillhouse at Ord Distillery, Distillers Company Ltd

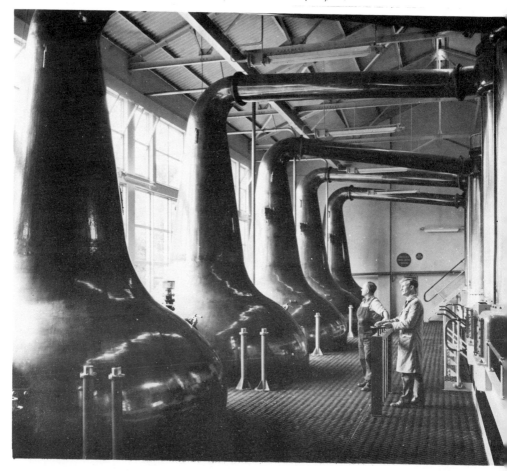

2.5 Chromatography

In Experiment 2.4, you found that ink is not a single substance. Ink can be separated by distillation into water and a pigment. It is interesting to find out whether the pigment is a pure substance or whether it can be split up into a mixture of substances.

You can try the method of *chromatography*. In this method you separate substances by passing their solutions through solids such as chalk or paper. It was discovered by a Russian chemist called Tswett. He worked chiefly on the extraction of coloured pigments. These gave the process the name of chromatography since *chromos* is Greek for *colour.* The reason why the substances separate is that each substance travels through paper or chalk at its own pace. If one substance is chemically attracted to chalk or paper, this will slow it down. If one substance is more soluble than others in the solution, it will tend to move quickly with the solvent and not stick to the paper. Travelling at different speeds, like competitors in a race, the substances become spread out, one in front and others spaced out behind.

Experiments 2.6 and 2.7 tell you how to try chromatography on black ink, using water as the solvent and paper or chalk as the separating solid. In Experiment 2.8, the solvent creeps up through the paper, and the method is therefore called ascending paper chromatography. You can use it to test the dyes in felt-tip pens, the colouring matter in Smarties®, and any food colouring matter that you can find at home. You will find that many of these dyes are mixtures of pigments.

One colour you see a good deal of around you is the colour green. It may occur to you to wonder whether grass and leaves are made of a green substance or whether they contain a green substance which can be separated from them. If they do, this green substance certainly does not wash out in the rain! We shall have to try solvents other than water to extract it from plants. Two common solvents are propanone and ethanol. Methylated spirit is an impure form of ethanol which you are likely to have available, and Experiment 2.9 uses this solvent to enable you to try chromatography on grass and leaves.

To find out whether the pigment in black ink is a pure substance

1. Carefully put a spot of black ink at the centre of a filter paper. Use a teat pipette as shown in Figure 2.10 (a).

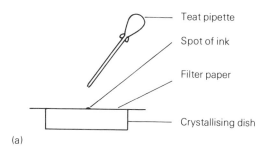

(a)

Figure 2.10

2. Place the filter paper on a glass dish and carefully add one drop of water to the spot, using a clean teat pipette Figure 2.10 (b).

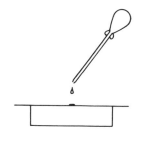

(b)

3. Wait until the water has stopped spreading, and then add a further drop of water.

4. Repeat step 3 until you are satisfied that the change you are watching is complete. It is most important to add the water to the centre of the spot and not to add it quickly. Be patient!

5. Make a note of everything you see as you add the drops of water.

6. Dry the filter paper in the oven. Keep it to staple into your book.

Now try the method shown in Figure 2.11 (a).

Figure 2.11 (a)

1. Make two cuts from the edge of a filter paper to the centre to produce a strip 6 mm wide. Bend it down so that the strip dips into the water when placed in the dish, as shown in Figure 2.11 (b).

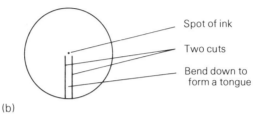

(b)

2. Carefully put a spot of black ink at the centre of the filter paper.

3. Wait. Make a note of everything you see.

4. When you are satisfied that the change you are observing is complete, dry the filter paper. Keep it to staple into your book.

Experiment 2.7

To find out whether chalk will separate the pigments in black ink

1. Take a piece of playchalk, not the dustless kind, and dab one end quickly into some black ink.

2. Stand the chalk on the inked end in a petri dish containing a small amount of water (see Figure 2.12). Watch what happens when the ink rises up through the pores in the chalk.

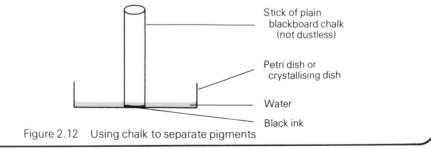

Figure 2.12   Using chalk to separate pigments

24

To investigate the pigments in felt-tip pens by means of paper chromatography

1. Take a strip of chromatography paper 12 cm × 6 cm, and rule a pencil line 2 cm from the short side. Put two spots of felt-tip pen on the pencil line, 2 cm apart, and write the colour in pencil under each spot. Write your name in pencil at the bottom (see Figure 2.13 (a)).

2. With the ink spots at the bottom end, turn over the top 2 cm of the paper and let this end hang over a glass rod. Fasten it with a paper clip.

3. Lodge the glass rod across the top of a 250 cm³ beaker containing water to a depth of 1 cm. The ink spots will be above the level of the water (see Figure 2.13 (b)). It is important that the paper does not sag against the side of the beaker. The purpose of suspending it over the glass rod is to prevent this. Water will travel up the paper to meet the spots.

4. When the water has risen nearly to the paper clip, take the piece of chromatography paper and put it into the oven at 50 °C to dry. Has the pigment separated into different components?

5. Repeat with different coloured felt-tip pens.

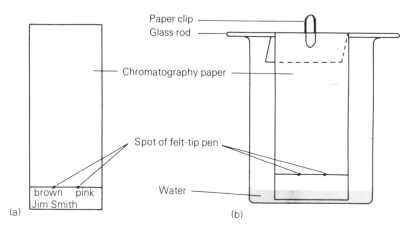

Figure 2.13   Ascending paper chromatography

To extract the green colouring from green leaves and try to separate it into components

1. Cut up a handful of privet leaves into very small pieces, using scissors. Alternatively, spinach leaves, grass, nettle, or any other green leaves can be used.

2. Put the pieces into a mortar with 3 cm³ methylated spirit and grind with a pestle until you have a really green extract.

3. Put a spot of this extract, using a teat pipette, onto a strip of chromatography paper 10 cm × 2 cm at a distance 2 cm from a short end. When the spot has dried, add another drop in exactly the same place. Apply five drops, one at a time (see Figure 2.14 (a)).

4. Put 5 cm³ methylated spirit into a boiling tube. Hold it in your hand to vaporise the methylated spirit and drive air out of the boiling tube. Stopper the tube with a rubber bung which carries a piece of wire bent into the shape of a hook.

5. Now hook the piece of chromatography paper bearing the green extract. Suspend it inside the boiling tube, with the green spot above the surface of the methylated spirit and the bottom edge of the paper just underneath the surface.

6. Wait until the methylated spirit has risen up the paper to the hook. Take the paper out, and hang it up to dry. Examine your result.

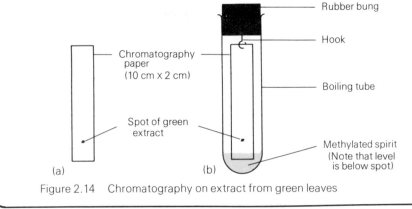

Figure 2.14   Chromatography on extract from green leaves

Questions on Chapter 2

1. You are given a mixture of powdered limestone, which is insoluble in water, and powdered washing soda, which is soluble in water. Describe what you would do to obtain both substances separately.

   filter it.

2. To obtain a pure liquid from a solution, what process would you use? Draw an apparatus which could be employed for this purpose.

3. When salt dissolves in water, has a physical change or a chemical change occurred? How can you prove that what you say is correct?

4. You are given a bottle of a red liquid used for colouring food. How can you find out whether the colouring material is one pigment or a mixture of pigments? Use diagrams in your answer.

5. On opening a bottle containing the white solid sodium iodide, you notice some black specks in it. Someone suggests that the black specks are iodine. How can you find out whether this is true? If the black substance is iodine, describe how you could separate it from the white solid. Draw the apparatus you would use.

6. The sweets called Smarties® are of different colours. Describe the steps you would have to take to find out whether the colouring materials used are pure pigments of mixtures of pigments. Draw the apparatus you would use.

Trace this grid on to a piece of paper, and then fill in the answers.

Crossword on Chapter 2

Across

2 An impure form of 8 across (4)
4 Material used in 3 down (5)
7 Method for separating miscible liquids (12)
8 Obtained by evaporating seawater (4)
11 Evaporate the first part of 12 down to obtain this (5)
12 Used for timing experiments (5)
14 A continent (4)
17 Patience is one (6)
18 Used with 5 down (6)
19 If you wear a lab coat, your clothes will not get these (6)
21 Obtained from grass by 3 down (11)

Down

1 Iodine _____ when heated (8)
2 This forms on iron (4)
3 Method used for separating pigments (14)
5 Used for crushing lumps (6)
6 The purest form of natural water (4)
7 Disperses through a solvent (9)
9 Where it is _____ (2)
10 Left behind after 7 across of crude oil (3)
12 Make sure this is off when you leave the lab (5, 3)
13 This is tied back in laboratories (4)
15 A laboratory is an _____ for working in (4)
16 A help in filtering (6)
20 Obtained by 7 across of crude petroleum (3)

3. Acids and alkalis

Some substances, such as vinegar and fruit juices, are said to have an *acid* taste. Chemists looked at these substances and analysed them to find out what they had in common. They found that all acids contain hydrogen, and they can all be made to give hydrogen gas in a chemical reaction. The acid in vinegar is *ethanoic acid* (often called *acetic acid*). The acid in fruit juice is *citric acid*. Other acids you have seen in the laboratory are *hydrochloric acid*, *nitric acid* and *sulphuric acid*. These acids also have an acid taste, but do *not* drink them because they are stronger acids than the ethanoic acid in vinegar!

Alkalis are substances like sodium hydroxide which have a soapy feel to the skin. The reason for this is that they convert the oil in the skin into soap. Alkalis are the opposite of acids in many chemical reactions. An alkali will cancel out an acid in a chemical reaction, and we call this *neutralisation*. The type of indigestion we call acid indigestion is caused by too much acid in the stomach. It can be cured by drinking a weak alkali to cancel or neutralise the extra acid. Table 3.1 gives a list of some common acids and alkalis and their uses.

It is useful to be able to tell whether a chemical is an acid or an alkali or neither, in which case it is said to be a *neutral substance*. A substance which will tell you whether a solution is acidic or alkaline or neutral is called an *indicator*. Many indicators can be extracted from plants. Experiment 3.1 tells you how this can be done, Experiment 3.2 uses some manufactured indicators, and Experiment 3.3 is about an indicator called universal indicator. Being a mixture of dyes, it turns a number of different colours in different solutions. It has different colours for strongly acidic and weakly acidic solutions, and it has different colours for strongly alkaline and weakly alkaline solutions.

Experiment 3.4 will enable you to find out what happens when an acid neutralises an alkali. Acids and alkalis are two big sets of chemicals, with thousands of members. The reaction between them, neutralisation, is one which you meet many times in Chemistry.

Table 3.1 *Some common acids and alkalis*

Name	Where you find it
Acids	
Citric acid	Fruit juices
Ethanoic acid (Acetic acid)	Vinegar
Carbonic acid	Fizzy drinks
Sulphuric acid	Car batteries
Alkalis	
Calcium hydroxide (Its solution is called limewater.)	Used for treating soil which is too acid to be fertile
Ammonia	Cleaning fluids
Magnesium hydroxide or *milk of magnesia*	Indigestion tablets and laxatives
Sodium hydroxide or *caustic soda*	Oven cleaners

3.1 Extraction of Indicators

Experiment 3.1

To extract an indicator from red cabbage and test it

1. Take two leaves of red cabbage. Tear them into pieces and put into a 250 cm^3 Pyrex beaker.

2. Cover with water. Warm and stir with a glass rod. The water will extract most of the colouring material to become a purple colour, and the leaves turn almost white.

3. Pour off the extract into a 50 cm^3 beaker.

4. Set out a row of test tubes in a rack. Fill each a quarter full with a different solution. Label the test tubes as you do so. Use solutions of ethanoic acid, hydrochloric acid, sodium hydroxide, sodium carbonate, and distilled water.

5. Using a teat pipette, drop three drops of the coloured extract into each test tube.

6. Make a note of the colour in each test tube. Make a table of results to show the name of the solution and the colour of the extract. From your results, you will see that the extract turns different colours in different solutions. It is purple in distilled water: this is its neutral colour. What is its acid colour? What is its alkaline colour (in sodium hydroxide solution)? The colouring matter in red cabbage is an indicator. It tells you whether solutions are acidic or alkaline or neutral.

7. Use the red cabbage extract to tell you whether the following solutions are acidic or alkaline or neutral: calcium hydroxide solution (limewater), sodium hydrogencarbonate solution, aluminium sulphate solution, and sodium chloride solution.

8. The preparation of an extract of indicator can be repeated with beetroot, elderberries, blackcurrants, blackberries, rose petals and many other fruits and petals of red, blue or purple hue.

3.2 Litmus and other indicators

Experiment 3.2

To find the acid, alkaline and neutral colours of indicators

1. Take a rack of clean test tubes. Fill the tubes quarter full with hydrochloric acid, sulphuric acid, distilled water, sodium hydroxide solution, calcium hydroxide solution. Label the test tubes as to their contents.

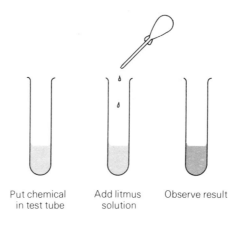

Put chemical   Add litmus   Observe result
in test tube   solution

2. With a teat pipette, add two drops of litmus solution. Record the colour in each solution.

3. Repeat the experiment with the indicators methyl orange and phenolphthalein.

4. Tabulate the colours of the three indicators, as shown in Table 3.2.

Table 3.2   *Colours of indicators*

Indicator	Neutral	Acidic	Alkaline
Litmus			
Methyl orange			
Phenolphthalein			

To find the pH numbers of some solutions, using universal indicator

1. Take a rack of clean test tubes. Fill them half full with the following solutions, labelling them as you do: sulphuric acid, hydrochloric acid, nitric acid, citric acid, ethanoic acid, distilled water, sodium hydrogencarbonate, sodium carbonate, ammonia solution, calcium hydroxide, sodium hydroxide.

2. Add three drops of universal indicator to each test tube.

3. In a table, write the name of each chemical. In the next column, write the colour of the indicator. In the third column, write the pH number which you have obtained by matching the colour of your solution with one of the shades on the universal indicator colour chart.

Table 3.3 *Universal Indicator Colours*

Chemical	Colour observed	pH number

3.3 Neutralisation

Experiment 3.4

> ### To find out what is formed when sodium hydroxide is neutralised by hydrochloric acid
>
> 1. With a measuring cylinder, measure 25 cm³ sodium hydroxide solution into a clean conical flask (see Figure 3.1 (a)).
>
>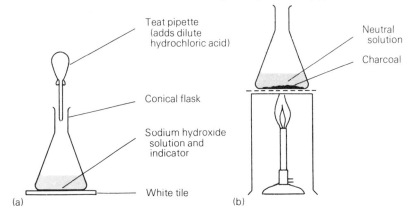
>
> Figure 3.1   Neutralisation of an alkali by an acid
>
> 2. Add three drops *(no more)* of screened methyl orange indicator.
>
> 3. Take dilute hydrochloric acid from a plastic 50 cm³ beaker with a teat pipette. Add the acid drop by drop, swirling the conical flask with your other hand as you do so. Stop when the neutral colour of grey is seen. If the solution goes purple, you have added too much acid and must repeat the experiment.
>
> 4. The solution is now neutral, but the indicator must be removed. Add a spatula measure of animal charcoal.
>
> 5. Warm and swirl the conical flask for five minutes (see Figure 3.1 (b)).
>
> 6. Filter. Evaporate the filtrate to dryness in an evaporating basin.
>
> 7. As an exception to our rules for laboratory procedure, in this experiment you can taste the product, carefully, on the end of a clean finger.
>
> 8. You can use other indicators for this experiment. You will see that the change from the alkaline colour to the neutral colour is easier to spot with some than with others.

3.4 Colours of Indicators

Table 3.4 gives the colours of the indicators you used in Experiments 3.1, 3.2, and 3.4.

Table 3.4   *Colours of indicators*

Indicator	Neutral solution	Acidic solution	Alkaline solution
Cabbage extract	Purple	Red	Green
Litmus	Purple	Red	Blue
Methyl orange	Orange	Red	Yellow
Phenolphthalein	Colourless	Colourless	Pink
Screened methyl orange	Grey	Purple	Orange in strongly alkaline solution; green in weakly alkaline solution

The results you have obtained with universal indicator in Experiment 3.3 will have shown you that this indicator has a number of colours. It is a mixture of dyes. It has a different colour for a strong acid and a weak acid, and for a strong alkali and a weak alkali. It turns red in hydrochloric acid, but in ethanoic acid it turns orange, and in citric acid it turns an orange-yellow colour. The strong alkalis, such as sodium hydroxide, turn the indicator violet, but weak alkalis, such as sodium hydrogencarbonate, turn the indicator blue. Sometimes, people cannot agree how to describe a colour. Some may describe a colour as indigo, when to others it looks violet. This is why, for universal indicator, each shade of colour is identified by a number, called the *pH number*. What you do is to look at the colour of the indicator in a solution, and match it with a chart showing universal indicator colours and their corresponding pH numbers. Then you can write down the pH number of the solution. You will see that acids have pH numbers less than 7, alkalis have pH numbers greater than 7, and distilled water – which is neutral – has a pH of 7. A strong acid has a lower pH number than a weak acid. A strong alkali has a higher pH number than a weak alkali.

We can draw a table:

Table 3.5   *Universal indicator colours and pH numbers in different solutions*

	Strongly acidic	Weakly acidic	Neutral	Weakly alkaline	Strongly alkaline
pH number	1 2 3	4 5 6	7	8 9 10	11 12 13
colour	Red Orange	Yellow	Green Blue	Indigo Violet	

By means of universal indicator, we can say whether a solution is acidic or alkaline, whether it is weakly or strongly acidic, and represent all this information simply by a pH number.

You may be thinking that pH is rather a peculiar term. It will be explained later in your course.

When one adds an alkali to an acid, the solution gradually becomes less acidic, until a neutral solution is formed. After this, if one goes on adding alkali, the solution becomes alkaline. When one adds an acid to an alkali, the solution becomes less alkaline and then neutral. If enough acid is added, the solution becomes acidic. The acid and alkali are acting like opposites, counteracting each other or, as we say, neutralising each other.

If you have done Experiment 3.4, neutralising hydrochloric acid with sodium hydroxide solution, the product you have obtained is a white crystalline solid, which tastes salty. In fact, it is common salt, sodium chloride. A chemical reaction has occurred with the formation of a new substance, which is different from the starting materials:

$$\text{Sodium hydroxide} + \text{Hydrochloric acid} \rightarrow \text{Sodium chloride} + \text{Water}$$

The arrow stands for 'form'. The reaction:

$$\text{Acid} + \text{Alkali} \rightarrow \text{Salt} + \text{Water}$$

is one which we shall meet many times.

Questions on Chapter 3

1. Write down the correct words to fill the blanks in these sentences:

    (a) Litmus turns _red_ in acids and _blue_ in alkalis.

    (b) A substance which tells us whether a chemical is an acid or an alkali is called an _indicator_.

    (c) A solution which is neither acid nor alkaline is said to be _neutral_.

    (d) Lemons contain _citric_ acid.

    (e) Fizzy drinks contain _carbonic_ acid.

    (f) 'Acid stomach' can be cured by drinking a solution of a weak _alkali_.

    (g) Strong oven cleaners contain the chemical _sodium hydroxide_.

    (h) Car batteries contain _sulphuric acid_ acid.

2. Someone tells you that the purple dye in elderberries is an indicator. Describe all that you would do to obtain the colouring material and to test it to find out whether this is true.

3. Take home some litmus paper. Use it to test foods and cleaning materials and anything else which interests you. Write down a list of acids and alkalis in your home.

4. What is the chemical name for common salt? Name an acid and an alkali which you can use to make common salt. Describe how you would make crystals of common salt from the acid and alkali you have named. _Sodium chloride_

Trace this grid on to a piece of paper, and then fill in the answers.

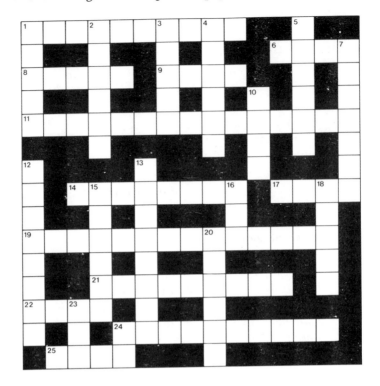

Crossword on Chapter 3

Across

1 A good source of indicator (3, 7)
6 It turns 15 down red (4)
8 Vinegar has an acid _____ (5)
9 Do this with books (4)
11 A member of 14 across (6, 9)
14 The opposite of acidic (8)
17 Make one of everything you see in an experiment (4)
19 This happens when 14 across react with an acid (14)
21 The name of one of 24 across (9)
22 Hotels (4)
24 These tell you whether a solution is acidic or alkaline (10)
25 The colour of 15 down when in 14 across (4)

Down

1 Those of some reactions are faster than those of others (5)
2 An acid in fruit juice (6)
3 Your beginning (5)
4 They are used for protection (6)
5 The acid in vinegar (6)
7 Protection from enemies (7)
10 Caustic _____ is another name for 11 across (4)
12 Cures acid stomach (8)
13 Another name for school holiday (8)
15 A member of 24 across (6)
16 This is an age (3)
18 Used to hold hot objects (5)
20 Laboratory rules are this (6)
23 Nothing (3)
24 Short for 'that is' (1, 1)

4. Elements and compounds

4.1 Metallic and non-metallic elements

The word *element* means *a pure substance which cannot be separated into simpler substances by any known chemical means.* Copper is an element: whatever you do with copper you cannot obtain anything simpler than copper from it. You can only build it into more complex substances like copper oxide and copper sulphate. These substances are *compounds*. Common salt is a pure substance. The methods of separation in Chapter 2 fail to split it into component parts, and prove that it is not a mixture. However, if common salt is melted and an electric current passed through it, it can be split up into sodium and chlorine, so it must be a compound of sodium and chlorine. Salt is not a mixture of sodium, which is a silvery grey solid, and chlorine, which is a poisonous green gas. It is a compound in which sodium and chlorine are chemically joined together and very difficult to separate. Sodium chloride is a white crystalline solid, quite different in appearance and behaviour from the elements which combined to make it. *A compound is a pure substance which consists of two or more elements chemically combined together.*

There are 92 elements found in nature. Some of the well known ones are listed in Table 4.1. Most of the elements fall into two groups, the metallic elements and the non-metallic elements. The elements copper, gold and iron are metals. Metals are *dense* and, when freshly cut, are *shiny*. They are *malleable* (which means that they can be hammered into shape) and *ductile* (they can be drawn out into wire). They are *sonorous*, which means they make a ringing noise when struck. Some non-metallic elements are carbon (which you know in the form of charcoal), chlorine (the gas used to disinfect swimming pools), and iodine (a black solid present in the antiseptic called tincture of iodine). Iodine has already been used in Experiments 1.2 and 2.3. These non-metals are less dense than metals and one (chlorine) is a gas with a much lower density. If you try to bend solid, non-metallic elements, they break. In general, non-metallic elements have a dull surface. Diamond is an exception. Many non-metallic elements are gases. Metals conduct heat better than non-metallic elements. It is almost true to say that metallic elements conduct electricity and non-metallic elements do not conduct. There is one exception, which you will discover in Experiment 4.1.

Table 4.1 *Some of the elements, with their symbols*

Element	Symbol	Element	Symbol
Aluminium	Al	Lead	Pb
Argon	Ar	Magnesium	Mg
Arsenic	As	Manganese	Mn
Barium	Ba	Neon	Ne
Calcium	Ca	Nitrogen	N
Carbon	C	Oxygen	O
Chlorine	Cl	Phosphorus	P
Chromium	Cr	Potassium	K
Copper	Cu	Radium	Ra
Gold	Au	Sodium	Na
Helium	He	Sulphur	S
Hydrogen	H	Tin	Sn
Iodine	I	Uranium	U
Iron	Fe	Zinc	Zn

For an explanation of symbols, see Chapter 7.

Table 4.2 *Characteristics of some elements*

Element	Colour	Surface, shiny or dull	Lump or powder	Can the shape be changed by hammering?	Does it conduct electricity?	Metallic or non-metallic element?
Carbon (charcoal)						
Carbon (graphite)						
Magnesium						
Lead						

To observe the characteristics of a number of elements

1. Make out an enlarged version of Table 4.2

2. Fill in your observations on these elements. Add aluminium, sulphur, iodine, copper, iron, calcium and any other elements.

3. Ask your teacher to show you sodium and phosphorus.

4. You will need a circuit to test whether the elements conduct electricity. Connect a battery, a switch and a bulb as shown in Figure 4.1, leaving a gap in the circuit. Attach crocodile clips to the wire on each side of the gap.

5. Close the gap with, say, a piece of copper, gripping it with the crocodile clips. Depress the switch and notice whether the bulb lights. If it does, a current of electricity must be flowing round the circuit, and the piece of copper closing the gap must be a conductor of electricity.

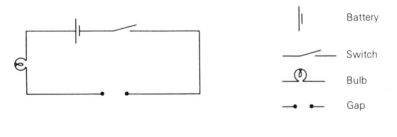

Figure 4.1   Circuit for testing conductors

6. Finally, enter into the last column of the table your opinion as to whether the element is metallic or non-metallic.

In Experiment 4.1, you made observations on a number of elements and tried to decide whether they are metallic or non-metallic. The differences are listed in Table 4.3.

Table 4.3   *Characteristics of metallic and non-metallic elements*

Metallic elements	Non-metallic elements
Solids (except mercury)	Some are gases, some liquids, some solids
Hard and dense	Most are softer than metals (except diamond)
Shiny when freshly cut	Dull (except diamond)
Malleable (can be hammered)	Easily broken when attempts are made to change the shape
Ductile (can be drawn into wire)	
Sonorous	Not sonorous
Conduct heat	Poor conductors of heat
Conduct electricity	Poor conductors of electricity or insulators (except graphite)

You will have found that there is one exception to the rule that metallic elements conduct electricity and non-metallic elements do not. Carbon is a non-metallic element which exists in different forms. Charcoal is one of its forms, and charcoal is a non-conductor. Graphite is another form of carbon, and yet it conducts electricity. It is shiny in appearance, but it does not have the strength characteristic of metals. As you handle it, flakes rub off on to your hands.

A substance can conduct electricity, and be a metal, without being an element. Brass and steel are metals, but they are not elements. Brass is an alloy, a sort of mixture, of the elements copper and zinc. Stainless steel is an alloy of the elements iron, carbon and chromium. A substance can be a non-conductor of electricity without being an element. The non-conductors, rubber and plastics, for example, are compounds.

Metallic and non-metallic elements are in some senses opposites. The saying that opposites attract one another seems to be true in Chemistry. Try these experiments to see what happens when you bring metallic and non-metallic elements into close contact.

4.2 Compounds

To bring together iodine and aluminium

1. Take half a spatula measure of aluminium powder. Put it into a test tube half full of water.

2. Grind some iodine in a mortar with a pestle. Add half a spatula measure to the aluminium. *Do not* grind aluminium and iodine together; this is dangerous.

3. Cork the test tube, and shake it from time to time over the next ten minutes.

4. Can you still see the silvery-grey colour of aluminium? Can you still see the black colour of iodine? Describe what you see. What do you think this substance is?

Demonstration Experiment 4.3

To grind together mercury and iodine

1. This experiment must be done in a fume cupboard by the teacher.

2. Take a metal or plastic tray. Place a mortar on the tray.

3. Take half a spatula measure of iodine, and put it into the mortar.

4. Place one small drop of mercury in the mortar.

5. Grind the mixture of mercury and iodine with a pestle. Be careful not to spill mercury as its vapour is poisonous. The tray is there in case you do spill any. Should you spill any, put sulphur on to the mercury in the tray.

6. Mercury is the only metal which is a liquid at room temperature. It has a beautiful silvery sheen and used to be called quicksilver. What happens to this shiny liquid as you grind it with iodine? Can you see the black colour of iodine? Describe what you see, and explain what has happened.

Experiment 4.4

To heat a mixture of iron and sulphur

1. Take a mixture of iron filings (7 g) and powdered sulphur (4 g). Make sure they are thoroughly mixed. Do two tests on the mixture.

2. Sprinkle some of the mixture on to a piece of paper. Bring a bar magnet underneath the paper and move it towards one end of the paper as shown in Figure 4.2 (a). Repeat.

3. Put a spatula measure of the mixture into a test tube half full of water. With a thumb on top, shake the tube.

4. Make a note of whether (2) and (3) have separated the mixture into iron and sulphur, and, if so, why.

5. Put some of the mixture into a Pyrex test tube until it is one third full. Clamp it at the open end.

6. Heat with a Bunsen burner right at the end of the tube as shown in Figure 4.2 (b). When you notice the mixture glowing red, take away the Bunsen and watch what happens.

7. Allow the tube to cool. Tip out the contents into a mortar. If you cannot obtain the contents any other way, you will have to smash the test tube. You must first wrap it up in a paper towel to prevent bits of broken glass flying about. Grind the contents with a pestle.

8. Repeat steps (2) and (3) to test the contents of the test tube.

9. Record whether (2) and (3) separated iron and sulphur.

10. Observe the appearance of the material you obtained from the test tube. Does it look the same as before heating? Can you see particles of grey iron and yellow sulphur in the mixture?

11. Is observation (9) on the mixture after heating the same as observation (4) on the mixture before heating? If it is different, can you explain why?

12. What did the red glow that spread through the Pyrex tube after you took away the Bunsen burner tell you?

13. What do you think you took out of the test tube?

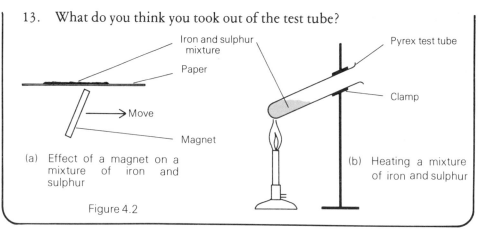

(a) Effect of a magnet on a mixture of iron and sulphur

(b) Heating a mixture of iron and sulphur

Figure 4.2

You have done three experiments on bringing together a metallic element and a non-metallic element. In Experiment 4.2, you saw the silvery grey colour of aluminium and the black colour of iodine fade and a new yellow solid form. This looked quite different from either element, and was in fact a compound of the two elements called aluminium iodide:

Aluminium + Iodine → Aluminium iodide

In Experiment 4.3, you saw the distinctive silvery liquid mercury and the black colour of iodine disappear. They were replaced by an orange solid. This is a compound of the two elements called mercury iodide:

Mercury + Iodine → Mercury iodide

In Experiment 4.4, you saw that a mixture of iron and sulphur could be separated by two methods: a magnet attracted the iron filings and left the sulphur behind; in water, iron sank and sulphur floated. On heating, a red glow spread through the mixture, suggesting that a chemical reaction was occurring. The product of heating looked different from the mixture. It was a grey solid with no specks of yellow sulphur in it, and it could not be separated as before into iron and sulphur. A chemical reaction had occurred to form a compound called iron sulphide:

Iron + Sulphur → Iron sulphide

In all these cases, two elements combine in a chemical reaction to form a new substance. The new substance is not a mixture of the two elements. It does not have the appearance or properties of the elements; it has a new appearance and new properties of its own; it is a compound of the two elements.

45

The differences between mixtures and compounds are summarised in Table 4.4.

Table 4.4 *The differences between compounds and mixtures*

Mixtures	Compounds
1 A mixture can be separated into two or more substances by physical methods.	A compound cannot be separated into other substances by physical methods.
2 A mixture has the properties of the substances present in it.	A compound does not have the properties of its elements.
3 No chemical change takes place when a mixture is made.	A compound is a new substance formed from its elements by a chemical reaction.
4 There is no heat given out or taken in when a mixture is prepared.	Heat is given out or taken in when a compound is formed.
5 A mixture can be prepared by mixing elements in any amounts, for example 1 g of sulphur and 99 g of iron or 99 g of sulphur and 1 g of iron.	A compound always contains its elements in fixed proportions, for example iron(II) sulphide contains 7 g of iron to 4 g of sulphur.

Questions on Chapter 4

1. What do you understand by the terms: (a) element, (b) compound, (c) mixture?

2. Name three elements and three compounds.

3. State whether the following are elements or mixtures or compounds: (a) steel, (b) lead, (c) sodium chloride, (d) charcoal, (e) brass, (f) copper sulphate.

4. Describe three differences between metallic elements and non-metallic elements. Name three elements which are metals and three elements which are non-metallic.

Trace this grid on to a piece of paper, and then fill in the answers.

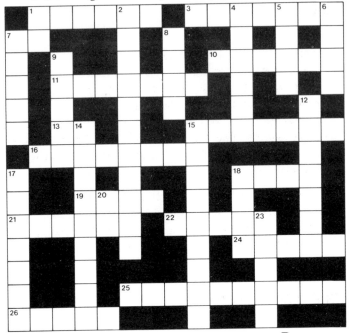

Crossword on Chapter 4

Across

1 This element can be drawn out into wire (6)
3 A simple form of matter (7)
7 Opposite of come (2)
10 An attractive alloy containing 1 across (6)
11 A liquid metal (7)
13 A contraction of senior (2)
15 This element is poisonous (7)
16 14 down These happen when elements combine (8, 9)
18 Often given out in 16 across, 14 down (4)
19 Don't do this in the exam! (4)
21 Group of six (6)
22 They soon spread in laboratories (5)
24 An alloy containing iron (5)
25 Nearly all non-metallic elements are _____ (they don't conduct 2 down) (10)
26 An alloy of copper and zinc (5)

Down

1 A short company (2)
2 Can be used to break up compounds (11)
4 Mistakes (6)
5 Abbreviation of electro-plated nickel silver (1, 1, 1, 1)
6 They are worn around the neck (4)
7 Polishing brings this on 10 across and 26 across (5)
8 Metallic elements are usually _____ (4)
9 Do not _____ apparatus. You may cut yourself! (5)
12 Not a compound but it may be turned to one by 16 across, 14 down (7)
14 See 16 across (9)
15 Metal used in manufacture of aircraft (9)
17 Do not _____ someone who is working (7)
18 Tools for the garden (4)
20 Religious education (1, 1)
23 Matter exists in one of three _____s (5)

5. Air

Air is vital to us. The most important thing in the world is getting our next breath. Air keeps us alive by means of a series of complex biochemical reactions. We are going to look at some simpler chemical reactions which take place in air and find out what part air plays in them. Air has always been regarded as an important part of science. The earliest chemists believed that the world was composed of four elements – earth, air, fire and water. We shall see whether they were correct in thinking air to be an element.

Perhaps you would like to try the experiments which follow to find out what happens when metals are heated in air. You will notice that certain changes occur, and you can do experiments to find out whether these changes would occur if you heated the metals in the absence of air.

5.1 Heating elements in air

Experiment 5.1

To heat metals in air

1. Take a piece of copper in tongs, and hold it in a Bunsen flame (gas half on, air hole half open) for five minutes. Lay down the copper on a heat-proof mat, and note its appearance.

2. Repeat, using magnesium. Wear safety glasses. Do not stare at the flame.

3. Put a piece of porcelain on a gauze supported by a tripod. A piece of broken crucible or evaporating basin is suitable. Put a piece of tin on to the porcelain, and heat for five minutes. Observe closely.

4. Repeat, using zinc. Repeat, using lead.

5. Make a table of the metals heated and your observations.

6. Have all these metals gained a coating of something different on being heated? When you scrape away the coating, can you see the metal underneath? What colours are the coatings on copper, magnesium, tin, zinc and lead?

What is the black coating on copper? Is it a black form of copper or is it a new substance formed from copper and air? The next experiment enables you to find out whether air is needed in the formation of the black coating.

Experiment 5.2

Is air necessary for the formation of a black coating on copper?

There are various ways in which you can keep out the air. Some are more efficient than others.

1. Try heating a piece of copper in a crucible with a lid. The lid is not designed to be airtight, and some air may enter. Is there any difference between the result you obtain and copper heated in unlimited air?

2. Cover some copper turnings in the bottom of a crucible with a layer of sand. Heat for ten minutes. Cool, and inspect the copper turnings.

3. Wrap a piece of copper in aluminium foil, and heat it for five minutes by holding it with tongs in a Bunsen flame. Cool and inspect.

4. Half fill a test tube with water. Drop in a piece of copper. Heat for five minutes. The water keeps air away from the piece of copper.

5. To heat copper in a stream of gas, which is certain to drive out all the air, is a good method. The apparatus in Figure 5.1 will serve.

6. Take a hard glass test tube. Stopper it, and heat at the closed end
(a) steadily at one spot for half a minute. The glass bursts under the pressure of hot gas, and you now have a test tube with a hole in the end to set up as in Figure 5.1. Connect one gas tap to A and connect another to a Bunsen burner at C. Put a piece of copper into the test tube.

(b) Turn on the gas at A, in the half-way position. Count five. Standing well back to avoid singeing your hair, light the gas at B. Turn down the gas until the flame is 3 cm high.

(c) Light the Bunsen at C and heat for five minutes.

(d) Switch off the gas at C. Keep the gas flowing at A until the copper is cool. Then switch off.

(e) Note whether a black coating has been formed.

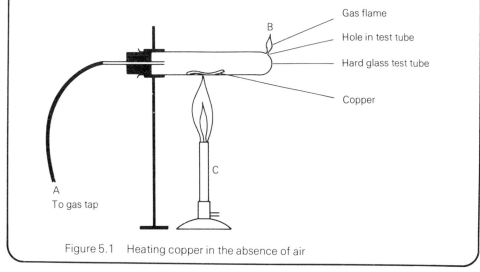

Figure 5.1   Heating copper in the absence of air

Experiment 5.3

Is there a change in mass when copper gains a black coating?

1. Fill a crucible one-third full of copper turnings. These must not be greasy as a loss of grease on heating will upset the results.

2. Put the crucible on to a top-loading balance, and read off the mass.

3. Put the crucible into a pipeclay triangle on a tripod (as in Figure 1.3) and heat it for about ten minutes, until the copper turnings look black. Do not use a lid. Cool.

4. Find the mass of the crucible and black copper turnings.

5. Subtract the smaller mass from the greater. Has there been an increase or a decrease or no change in mass? Compare results with the rest of your class. Have you been able to decide whether copper gives something to or takes something from the air or does neither? Since this is an important turning point in our work on air, you may like to check your result by means of another experiment along the same lines.

Is there a change in mass when magnesium burns to form a white solid?

1. Fill a crucible one third full with magnesium ribbon. Put on the lid.

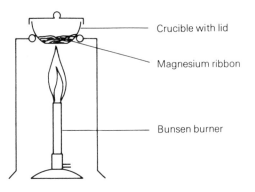

Figure 5.2 Heating magnesium in a crucible

2. With a top-loading balance, find the mass of the crucible and magnesium and lid.

3. Heat the crucible on a pipeclay triangle. From time to time lift the lid with tongs to let air in. The lid prevents particles of white solid from blowing away.

4. When the magnesium no longer glows when you lift the crucible lid, the reaction is over. Switch off the gas, and allow the crucible to cool.

5. Find the mass of the crucible and lid and white solid.

6. Calculate whether there has been a change in mass. Compare results with the rest of your class. Have you decided whether magnesium gives something to the air or takes something from the air or does neither, when it forms a white solid?

You will have seen in Experiment 5.1 that metals become coated with a different substance when heated in air. A black coat forms on copper, a white coat forms on magnesium, a yellow coat forms on lead, and zinc acquires a coat which is yellow when hot and white when cold. In order to explain what the coat on each metal is, we have to know whether air is involved in this reaction. If the reaction is a change from one form of, say, copper to another form of copper, the reaction will take place in the absence of air.

Some of the methods of excluding air which you tried in Experiment 5.2 work better than others. You will have found that, when air is carefully excluded, copper does not gain a black coat on heating. This opens up three possibilities. When copper gains a black coat, copper

(a) gains something from the air, or

(b) gives something to the air, or

(c) neither gives to nor takes from the air.

A little thought will tell you how to decide between the three possibilities. An experiment will tell you whether there is

(a) an increase in mass, or

(b) a decrease in mass, or

(c) no change in mass when copper gains its black coat.

If you have done Experiments 5.3 and 5.4, you will have found out that there is an increase in mass. This proves that copper takes something from the air to form its black coat, and magnesium takes something from the air when it forms a white solid.

These experiments lead us to think about whether all the air is able to combine with copper and with magnesium or whether only part of the air is used up. Demonstration Experiment 5.5 enables us to find out.

There are many substances which react with air. So far, we have studied metals. Experiment 5.6 will tell you whether a candle uses air when it burns. This experiment was done 300 years ago by John Mayow as part of the early research on air. Phosphorus is a very reactive non-metallic element, and Demonstration Experiment 5.7 will show you whether phosphorus uses up part of the air when it burns.

5.2 Experiments to find out what fraction of the air will take part in chemical reactions

To find the fraction of air used up when copper gains its black coating

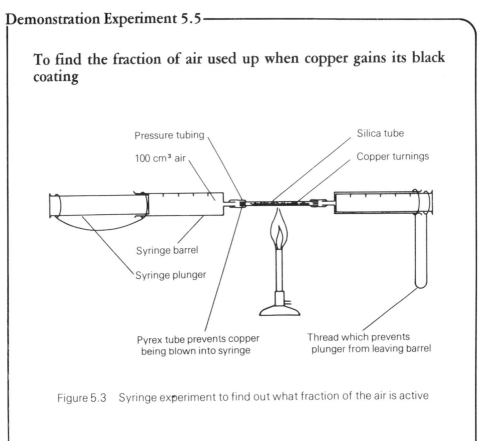

Pressure tubing

100 cm³ air

Silica tube

Copper turnings

Syringe barrel

Syringe plunger

Pyrex tube prevents copper
being blown into syringe

Thread which prevents
plunger from leaving barrel

Figure 5.3 Syringe experiment to find out what fraction of the air is active

1. Set up the apparatus shown in Figure 5.3. The glass syringes each have a capacity of 100 cm³. Strings hold the plungers to the barrels in case a plunger should fall out. The syringes are both clamped at the same height and connected by pieces of pressure tubing to a narrow silica tube containing copper turnings. The connections must be airtight. At the beginning of the experiment, have 100 cm³ of air in one syringe and no air in the other.

2. Heat the tube containing copper, and drive the air slowly through it from one syringe to the other. Then push the air slowly back. After three minutes, stop heating and allow the apparatus to cool. Measure the volume of air.

3. Repeat the heating, pushing air slowly to and fro from one syringe to the other. Cool again, and measure the volume again.

4. A time will come when the volume of air stops changing. When it has reached a steady value, note the volume of air left in the syringe.

5. Test the gas in the syringe. Drive it into a test tube by pushing in the plunger. Take a lighted taper, and put it into the test tube. Does it continue to burn?

6. The active part of the air is the difference between the 100 cm³ of air you started with and the volume of air left.

Experiment 5.6

Is part of the air used up when a candle burns?

1. Attach a candle to a cork raft by melting the base of the candle. Float it in a trough of water.

2. Light the candle, and put a bell jar over it as shown in Figure 5.4 (a).

3. Stopper the bell jar and wait.

4. The experiment can be done with a gas jar instead of a bell jar.

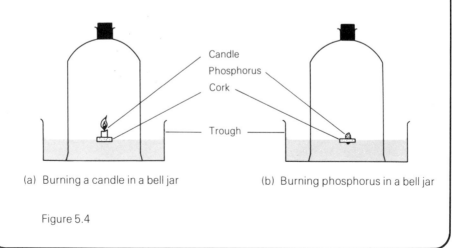

Candle
Phosphorus
Cork
Trough

(a) Burning a candle in a bell jar   (b) Burning phosphorus in a bell jar

Figure 5.4

Demonstration Experiment 5.7

Is air used up when phosphorus burns?

1. Phosphorus is a non-metallic element. It is dangerous to handle: it bursts into flame in the air, and is kept under water. Carefully cut a small piece of phosphorus away from a lump of the element. Hold the lump of phosphorus with tongs and cut it under water. Attach the small piece of phosphorus to the point of a pin sticking through a cork raft.

2. Float the phosphorus in a trough and cover it with a bell jar as shown in Figure 5.4 (b).

3. Heat the end of a piece of wire and poke the wire through the mouth of the bell jar to touch the phosphorus.

4. Quickly remove the wire and stopper the bell jar.

5. Note the colour of the flame and the appearance of the product formed when phosphorus burns. Notice what happens to the water level in the bell jar.

Let us see what the results of all these experiments have told us about the air. In Demonstration Experiment 5.5, you saw an accurate determination of the fraction of air which combines with copper. Your result should be that, if you start with a syringe containing 100 cm^3 of air, at the end of the experiment there will be 79 cm^3 left. This means that 21 cm^3 out of 100 cm^3 of air are used by the copper; that is, 21% or about one fifth.

This experiment can be repeated with other metals, such as tin, zinc and iron. The result is always the same: one fifth of the air is active and four fifths of the air is inactive and will not react with metals.

John Mayow was one of the first people to do research on air. He was interested in burning because it is one of the most important everyday reactions which take place in air. In repeating one of the experiments he did (Experiment 5.6) you burned a candle. This is not an element but is a mixture of compounds obtained from crude petroleum oil. You saw the candle flame go out, and water rise up inside the bell jar to take the place of the air which had been used up. The gas left in the bell jar contains gases formed from the burning candle as well as the inactive part of the air. This is why we cannot use this experiment to tell us exactly what fraction of the air is used up.

On repeating the bell jar experiment with phosphorus, you saw that phosphorus burns fiercely with a yellow flame. The heat evolved causes the air inside the bell jar to expand and push the water level down. The flame goes out for lack of active air, and water rises up inside the bell jar to take the place of the air used up. The product formed is a fine white powder, which gradually settles to the bottom of the bell jar, and dissolves in the water.

Your experiments have shown that air is a mixture of at least two gases, an active gas, which makes up one fifth of the air, and an inactive gas which forms four fifths of the air. Before giving names to these gases, you may like to read about some famous chemists who worked out the composition of air many years ago.

5.3 Some early experiments on air

The early chemists did many experiments on air, with metals and with non-metallic elements. They found that it was difficult to explain their results and still go on believing that air was an element, but, since it had been believed for centuries that air was an element, they clung to this view. A turning point in the work on air came with a crucial experiment done by a British scientist called Joseph Priestley in 1774. Priestley did research in his spare time while working as a Unitarian minister.

Priestley took a solid called *red calx of mercury*, which had been made by heating mercury in air. He had the idea of heating the solid to a higher temperature, and he did this by focusing rays of sunlight on to it with a lens. He found that a gas was given off, and since he had become expert at collecting gases in glass apparatus by displacement of mercury, he collected the gas. Air was the only gas he knew, so he assumed that it was a type of air. He found that materials such as a taper and a piece of charcoal burned much more brightly in this kind of air than in what he called *common air*.

Priestley had done experiments like Mayow's (Experiment 5.6), floating a mouse in a cage on the cork instead of a candle. He found that the mouse used up part of the air only, and when it started gasping for breath, he pulled it out. There was air left in the bell jar, but not the kind of air that the mouse needed. He found that the mouse revived much faster in the new air he had discovered than in ordinary air. He estimated that he had 'procured air between five and six times as good as the best common air that I have ever met with'. Priestley tried breathing it himself, and found that it produced 'a very light, easy sensation in the chest'. He wrote, 'Who can tell but that in time this pure air may become a fashionable article in luxury? Hitherto only my mice and myself have had the breathing of it'. He realised, however, that we might live too fast in the pure air, and the *common air* was safer for us.

Although Priestley had discovered a new gas, he did not realise this, and continued to puzzle over his findings and try to explain them in terms of different kinds of air. He discussed his experiments with French scientists, including Antoine Lavoisier, a French aristocrat with great enthusiasm for doing chemical experiments in his spare time. Lavoisier realised that Priestley had discovered something momentous. He repeated Priestley's experiments.

Lavoisier took mercury in a sealed vessel in contact with a limited supply of air. He kept it just below its boiling point for twelve days, and watched red calx of mercury form on the surface. At the same time one fifth of the air in the reservoir was used up. Figure 5.5 shows how this was done. He then took the red solid and heated it in an apparatus which enabled him to measure the volume of 'air' evolved. He found that the volume of 'air' evolved when the red calx was heated was the same as the volume of air taken up when mercury was heated in air to form the red calx.

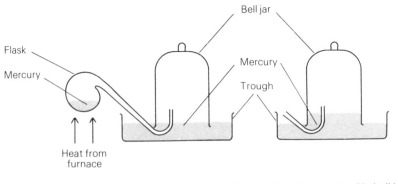

(a) Heating mercury in a limited supply of air   (b) Mercury level in bell jar

Figure 5.5   An illustration of Lavoisier's experiment

Lavoisier explained his results by breaking with tradition and rejecting the view that air was an element. He said that air was a mixture of an active gas, which he called *oxygen*, and an inactive gas which he called *azote*. Oxygen means *acid-maker*; Lavoisier chose this name because many elements, for example sulphur and carbon, combine with oxygen to form acids. Azote means without life: Lavoisier chose the name because the gas does not support life. We now call this gas *nitrogen*. Priestley is famous as the discoverer of oxygen. Lavoisier's explanation of their results was a very important discovery, because it made sense out of a whole series of experiments by different workers on combustion of metals and non-metallic elements. Combustion was seen to be combination with oxygen to form an oxide, and 'red calx of mercury' was seen to be mercury oxide. Lavoisier, for his masterful interpretation of the accumulation of years of work by different scientists in terms of a simple theory, combination with oxygen, is hailed as 'the founder of modern chemistry'.

Priestley's prediction came true. Oxygen, though not exactly an expensive luxury, is a gas which can be used by people who need help in breathing. Pneumonia patients, whose lungs cannot work properly, and traffic police, whose lungs have become full of polluted air by directing traffic at busy road intersections for long periods, are among those who use oxygen.

These famous scientists met with little respect in their lifetimes. In England, in 1794, Priestley was forced to leave the country because of his unorthodox religious views. During the French Revolution, in 1794, a council of revolutionary citizens accused Lavoisier of having misused public money. There was no foundation for the charge, but the council was determined to execute aristocrats and sent Lavoisier to the guillotine.

The reactions we have studied can be explained as combination with oxygen, a process called oxidation.

$$\text{Copper} + \text{Oxygen} \rightarrow \text{Copper oxide}$$

$$\text{Magnesium} + \text{Oxygen} \rightarrow \text{Magnesium oxide}$$

Burning is oxidation accompanied by a flame. For example, magnesium burns in oxygen to form magnesium oxide. *Combustion is oxidation in which heat energy is given out*. We talk about the combustion of food in the body; in this type of combustion there is no flame. In the combustion of paraffin in a stove, there is a flame, and we call this type of combustion *burning*.

5.4 Methods of preparing oxygen; burning elements in oxygen

There are various methods of preparing oxygen in the laboratory. You may find it interesting to repeat Priestley's experiment. If you heat mercury oxide, oxygen is driven off and little silvery balls of liquid mercury are left in the heated test tube. The reaction:

Mercury oxide → Mercury + Oxygen

has occurred. Since mercury vapour is poisonous, Demonstration Experiment 5.8 must be done in a fume cupboard.

There are compounds rich in oxygen and less expensive than mercury oxide which will give up their oxygen. The liquid, hydrogen peroxide, is one. It is a compound of oxygen with an element called hydrogen, and it is a compound which splits up fairly readily to give oxygen. If you take the stopper off a bottle of hydrogen peroxide, it will give oxygen; if you warm it, it will give oxygen faster. There are a number of substances which can be added to speed up the formation of oxygen from hydrogen peroxide without heating. They are called *catalysts*, and they are said to *catalyse* the reaction. A compound of manganese and oxygen called manganese(IV) oxide is the catalyst we are going to use in Experiment 5.9. (Chapter 7 contains an explanation of what (IV) means.)

You will need to be able to test the gas formed to see whether it is oxygen. Since air is only one fifth oxygen, materials burn five times as well in pure oxygen as they do in air. The usual test for oxygen is to put into a jar of oxygen a wooden splint which you have lit and then almost blown out – a splint which is still glowing. If the gas is oxygen, the splint burns brightly. *Oxygen rekindles a glowing splint.*

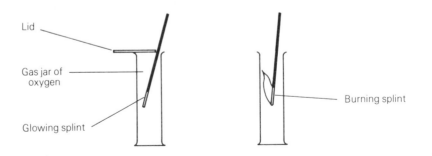

Lid

Gas jar of oxygen

Glowing splint

Burning splint

Figure 5.6   A glowing splint rekindled by oxygen

To prepare oxygen from mercury oxide

You may find it interesting to see Priestley's experiment repeated. Since mercury vapour is poisonous, this experiment must be done in a fume cupboard by the teacher.

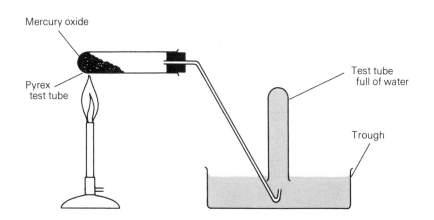

Figure 5.7   Preparation of oxygen from mercury oxide

1. Set up the apparatus shown in Figure 5.7 with red mercury oxide in a hard glass test tube. A delivery tube leads to a test tube full of water, inverted over a trough of water.

2. Heat the mercury oxide, and collect the gas evolved. Cork the test tube full of gas and put it into a rack.

3. Put a glowing splint into the test tube of gas.

4. You know from your reading that the gas is oxygen. What does it do to a glowing splint? What can you see in the heated test tube in addition to red mercury oxide?

To prepare oxygen and to burn some elements in it

1. Assemble the apparatus shown in Figure 5.8. Put a spatula measure of manganese(IV) oxide in the side-arm tube.

2. Lay four boiling tubes in the trough so that they fill with water. Hold one over the end of the delivery tube.

3. Add some hydrogen peroxide to the manganese(IV) oxide.

4. Discard the first tube of gas collected as this is largely air. As soon as the second tube is full, put a cork in the tube under water, and quickly place the open end of the next tube over the delivery tube.

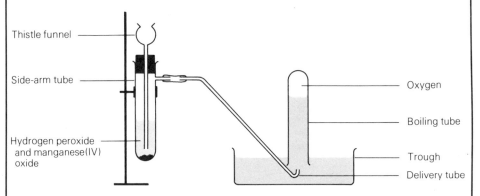

Thistle funnel

Side-arm tube

Hydrogen peroxide
and manganese(IV)
oxide

Oxygen

Boiling tube

Trough

Delivery tube

Figure 5.8   Apparatus for preparing oxygen

5. Collect and cork four boiling tubes of oxygen and stand them in a rack. Carry out the following tests, with the tubes standing in the rack (*not* in your hand). Do not take out the cork until the burning sample is ready to be put into the tube. After each test, add three drops of litmus solution, cork and shake the tube.

6. *Carbon*. Place some powdered carbon in a combustion spoon. Heat to redness, then quickly put the spoon into a boiling tube of oxygen.

7. *Sulphur*. Place some sulphur in a combustion spoon. Heat until the sulphur burns, and quickly transfer to the tube of oxygen.

8. *Iron*. Wrap some iron wool round a combustion spoon. Heat until it is red hot. Put into a tube of oxygen.

9. *Magnesium*. Wrap a short piece of magnesium ribbon round a combustion spoon. Ignite it, and quickly place in a tube of oxygen. Do not stare directly at the flame.

10. Ask your teacher to demonstrate the combustion of sodium, calcium and phosphorus in oxygen.

11. Make a table to show your observations on the colour of the flame, the nature of the product and the action of litmus on a solution of the product.

You have prepared oxygen and tested it with a glowing splint and burned a number of elements in it. The results of these experiments are very important. In all cases, you will have noticed that oxygen allows the elements to burn more brightly than air does. Oxygen is a better supporter of combustion than air is. Your table of results should look like this.

Table 5.1   *Combustion of elements in oxygen*

Element	Metallic or non-metallic	How does it burn?	Appearance of product	Colour of litmus in solution of product	Is solution acidic or alkaline?	Is product an acid or a base?
Carbon	Non-metal	Red glow	Colourless gas	Red	Acidic	Acid
Sulphur	Non-metal	Blue flame	Misty gas	Red	Acidic	Acid
Iron	Metal	Yellow sparks	Blue-black solid	Insoluble		
Magnesium	Metal	White light	White solid	Blue	Alkaline	Base
Sodium	Metal	Yellow flame	Yellowish solid	Blue	Alkaline	Base
Calcium	Metal	Red flame	White solid	Blue	Alkaline	Base
Phosphorus	Non-metal	Yellow flame	White solid	Red	Acidic	Acid

When elements burn in oxygen, they combine with oxygen to form oxides. A compound of oxygen and one other element is called an oxide. The oxide of carbon is an invisible gas. The oxide of sulphur is a fuming gas. These are fresh examples of the amazing differences you see between a compound and the elements which combine to make it. The oxides of iron, magnesium, sodium calcium and phosphorus are formed when these elements burn in oxygen. They are described in Table 5.1.

Metals give oxides which are bases. Some bases, for example iron oxide, are insoluble and cannot be tested with litmus. Soluble bases are called alkalis. Alkalis are a sub-set of bases. Bases are compounds which react with acids to give a salt and water. Non-metallic elements give oxides which are acidic. There are, however, some oxides of non-metallic elements which are neutral and in-soluble. It was because Lavoisier had done a large number of experiments on sulphur and phosphorus that he called the new gas acid-maker (or oxygen). Table 5.2 summarises the information you have obtained.

Table 5.2   *Products of burning elements in oxygen*

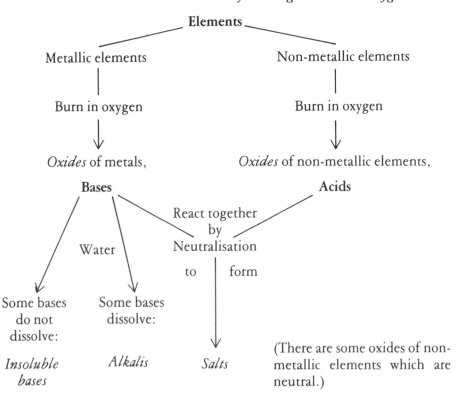

64

To obtain nitrogen from the air

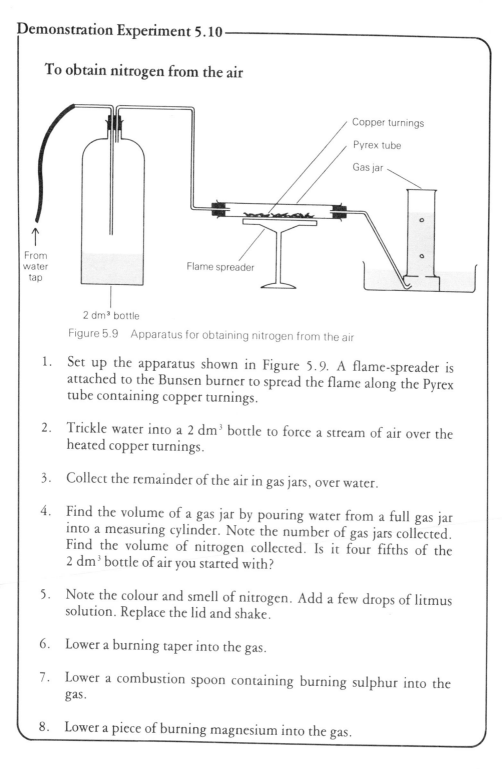

Figure 5.9   Apparatus for obtaining nitrogen from the air

1. Set up the apparatus shown in Figure 5.9. A flame-spreader is attached to the Bunsen burner to spread the flame along the Pyrex tube containing copper turnings.

2. Trickle water into a 2 dm³ bottle to force a stream of air over the heated copper turnings.

3. Collect the remainder of the air in gas jars, over water.

4. Find the volume of a gas jar by pouring water from a full gas jar into a measuring cylinder. Note the number of gas jars collected. Find the volume of nitrogen collected. Is it four fifths of the 2 dm³ bottle of air you started with?

5. Note the colour and smell of nitrogen. Add a few drops of litmus solution. Replace the lid and shake.

6. Lower a burning taper into the gas.

7. Lower a combustion spoon containing burning sulphur into the gas.

8. Lower a piece of burning magnesium into the gas.

5.5 Nitrogen

Nitrogen makes up 78% of the air. It is the inactive part of the air. Although nitrogen plays no part in the processes of breathing and burning you would be wrong to suppose that nitrogen is unimportant. A good supply of nitrogen compounds is essential for the growth of plants. If the soil is not rich in nitrogen compounds, it yields poor crops. Artificial fertilisers containing nitrogen compounds must be added to improve the soil.

Nitrogen can be obtained from the air by taking out the more reactive part of the air, oxygen. There are a number of ways of doing this. Experiment 5.10 uses heated copper to remove the oxygen. The nitrogen you obtain in this way contains small amounts of other gases, which are not removed by heated copper.

Nitrogen is a colourless, odourless, insoluble gas. It does not support combustion, except for the combustion of burning magnesium. Magnesium burns in nitrogen to form magnesium nitride, a compound of magnesium and nitrogen.

Oxygen and nitrogen from the air

In Experiment 2.5, we saw how the liquids water and ethanol could be separated by distillation. It is possible, though very difficult, to liquefy air and distil it to give both oxygen and nitrogen. Many gases can be liquefied by cooling and compressing them. This method does not work for air as a very low temperature must be reached before it will liquefy, and it is impossible to find anything cold enough to cool air down so far. The method discovered by Joule and Thompson is used. They found that when a gas is compressed and then suddenly allowed to expand, the expansion results in a cooling. By this method, air can be cooled sufficiently to liquefy it.

Liquid air is a transparent, pale blue liquid. It must be kept below –190 °C in a special type of thermos flask called a *Dewar flask*, which prevents heat from entering. Liquid air is distilled in a carefully insulated fractionating column (see Figure 5.10), where a slight adjustment in temperature results in the liquid boiling. The temperature is controlled to give only a few °C difference between the top and the bottom. The column is filled with perforated shelves at different levels. The ascending and descending liquids and gases are thoroughly mixed at each level. Since the top of the column is a few °C colder than the bottom, the liquid with the higher boiling point, oxygen, finds it more difficult to remain a gas at the top of the column than does nitrogen, which has a lower boiling point. Gradually oxygen accumulates at the bottom as a liquid, and the more easily vaporised nitrogen is taken off from the top as a gas. The boiling points are −196 °C for nitrogen and −183 °C for oxygen. The separation of nitrogen and oxygen from air by distillation is proof that air is a mixture, not a compound.

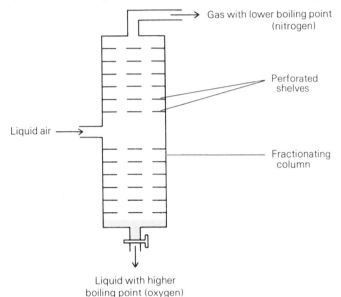

Figure 5.10   Distillation of liquid air

5.6 The composition of air

There are other gases present in the air in addition to oxygen and nitrogen. The composition of the air is shown diagrammatically in Figure 5.11. The nitrogen you made in Demonstration Experiment 5.10 actually contained noble gases and carbon dioxide and water vapour.

Oxygen

Pure oxygen is stored under pressure in strong metal cylinders. It has many uses.

Space rockets. In addition to fuel, space rockets also carry their own oxygen for the fuel to burn in, as there is no oxygen in space. The Saturn rockets, which were used to lift American astronauts into orbit for journeys to the moon, carried over 2200 tons of liquid oxygen. The first stage, while the jets were roaring, burned 450 tons of kerosene in 1800 tons of oxygen. Stages two and three were powered by hydrogen burning in oxygen. There was also some oxygen aboard for the astronauts to breathe.

Steel. One ton of pure oxygen is needed for every ten tons of steel produced from impure iron. The carbon in the impure iron burns away to form carbon dioxide. Modern steel works are usually equipped to make their own oxygen on the site.

Oxy-acetylene cutting and welding. Acetylene burns more vigorously in pure oxygen than in air, producing a flame with a temperature of about 3000 °C. This temperature will melt iron and steel, which can then be cut or welded (joined together). This is shown in Figure 5.12.

Hospitals. In hospitals, oxygen is used to help patients with breathing difficulties, such as pneumonia cases and asthmatics. It is used to revive people who have been dragged out of smoke-filled rooms and people rescued from drowning. It is mixed with anaesthetic gases during surgical operations.

Pollution. Oxgen is used to fight pollution. Figure 5.13 shows the River Protection Service cleaning up the environment by piping oxygen from a tank into polluted river water.

Mountaineers and divers. The air becomes less dense as you climb a mountain. On Mount Everest, which is five miles high, and on similar peaks, there is not enough air for mountaineers to breathe, and they have to carry cylinders of oxygen on their backs. High-altitude fliers also carry oxygen with them. Underwater explorers either carry oxygen in portable cylinders attached to breathing masks or work inside special capsules, which are supplied with oxygen.

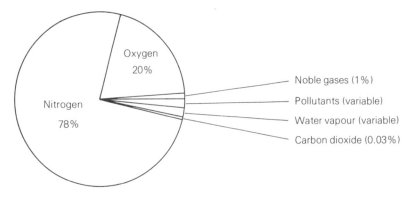

Figure 5.11   The composition of air

↑ Figure 5.12   An oxy-acetylene flame   ↓ Figure 5.13   Using oxygen to fight pollution

Nitrogen

Nitrogen is used in the manufacture of ammonia. This compound is a useful cleaning substance, and it has enormous use in the manufacture of fertilisers. From ammonia, another nitrogen compound called nitric acid is made. This too is important in the manufacture of fertilisers. Nitric acid is also used in the manufacture of explosives, such as TNT (trinitrotoluene).

Carbon dioxide

Carbon dioxide is only 0.03% of the air, but it is absolutely essential. Without carbon dioxide, plants could not grow; without plants, animals would starve; and without vegetable crops and meat to eat, we would not survive. Nature has a mechanism for ensuring that the level of carbon dioxide remains at 0.03%.

Water vapour

The content of water vapour in the air varies, from place to place and from time to time. On an average day in Britain, it might be 4%.

Noble gases

The noble gases are a group of gases which are elements and which are chemically unreactive. For a long time, it was believed that they formed no compounds at all, and they were called the *inert gases*. Recently, it was discovered that they can be made to combine with one or two other elements, under very drastic conditions. They are called *helium*, *neon*, *argon*, *krypton* and *xenon*. Helium is very light and is used in balloons sent up for meteorological purposes. Neon and argon are used in artificial lights.

Pollutants

The content of impurities in the air is greater in towns and industrial centres than in the country. Carbon monoxide is a poisonous gas which comes from the exhausts of motor vehicles. Sulphur dioxide is a poisonous gas which comes from the combustion of fuels containing sulphur. Hydrogen sulphide is an unpleasant-smelling and poisonous gas which pollutes the atmosphere.

It is possible to detect the carbon dioxide and water vapour in the air by using these tests:

(1) Carbon dioxide turns a solution called limewater milky (see Chapter 6.6).

(2) Water turns white anhydrous copper sulphate blue (see Chapter 1.4).

To test for carbon dioxide and water vapour in the air

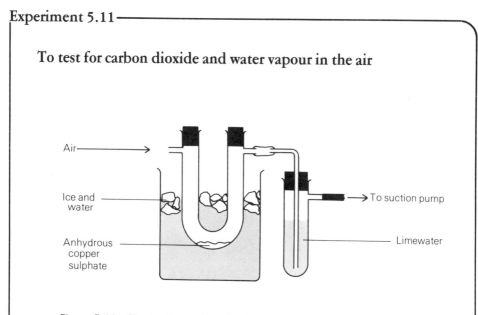

Figure 5.14   Testing for carbon dioxide and water vapour in air

1. Set up the apparatus as shown. The U tube stands in ice-water to condense any water vapour out of the air. It contains anhydrous copper sulphate. *Anhydrous* means *without water*. Anhydrous copper sulphate is a white powder. When it reacts with water, it forms a compound of copper sulphate and water, which is blue. The side-arm tube contains limewater. Limewater is a solution of calcium hydroxide.

2. Using a suction pump, draw a stream of air through the apparatus.

3. Watch to see whether the limewater turns milky.

4. Note whether a liquid condenses inside the U tube, and whether it turns the white anhydrous copper sulphate blue.

5.7 The rusting of iron

We began our study of air by looking at metals reacting with air. A very important reaction of a metal in air is the rusting of iron. Let us find out more about it. Experiments 5.12, 5.13, 5.14 are designed to allow you to find out several things about iron rusting.

Does iron increase in mass on rusting?
Does iron use up part of the air when it rusts?
Does iron use up the same fraction of the air as the metals which burn in air?
Does iron need both water and air to rust?
Do any other substances need to be present for iron to rust?
What chemical reaction takes place when iron rusts?
What is the chemical name for rust?

Experiment 5.12

Is there a change in mass when iron rusts?

1. Put some dry iron filings into a petri dish. Put the dish on to a top-loading balance, and read off its mass.

2. Add water carefully with a teat pipette so that the filings are wet but none are washed away.

3. Put the petri dish on a window sill and leave it for a week. The water should evaporate. If necessary, put the petri dish and iron filings into the oven to finish drying.

4. Since the water you added has now evaporated, the petri dish and iron filings should have the same mass as before. Reweigh to find out whether this is so.

5. You now know:

   Mass of: petri dish + iron filings in (1) = m_1 grams
   Mass of: petri dish + iron filings in (4) = m_2 grams

   Are the figures m_1 and m_2 the same? If not, has the mass increased or decreased? What change do you notice in the appearance of the iron filings?

Does iron combine with air on rusting?

1. Take a 100 cm³ graduated tube. To make the volume of air inside the tube 100 cm³, add some water and then turn the tube open side down with your thumb over the end. Carefully let water trickle out until you have 100 cm³ of air. Invert the tube in a deep vessel, and read the volume of air with the water levels inside and outside the tube equal, as shown in Figure 5.15 (a).

2. With your thumb over the open end, turn the graduated tube open end upwards, and add a spatula measure of iron filings. Carefully invert in a beaker of water so that the iron filings stick to the side of the tube.

3. Leave to stand for a week (see Figure 5.15 (b)).

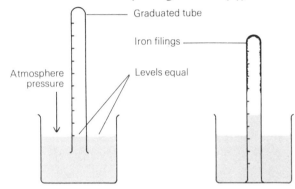

(a) Ensuring the tube contains 100 cm³ of air   (b) Iron rusting in graduated tube

Figure 5.15   Experiment to find what fraction of air combines with iron

4. Put your thumb over the open end, and then open the graduated tube in a deep vessel, so that you can read the volume of air at atmospheric pressure, with the water levels inside and outside the tube equal. Note the volume.

5. Test the gas in the graduated tube with a lighted splint.

6. Your results tell you:

$$\text{Initial volume of air} = 100 \text{ cm}^3$$

$$\text{Final volume of air} = v \text{ cm}^3$$

$$\text{Volume of air used up} = (100 - v) \text{ cm}^3$$

$$\text{Percentage of air used up} = (100 - v)\%$$

To investigate the conditions necessary for iron to rust

1. Set up conical flasks (or specimen tubes or test tubes) containing iron nails under different conditions, as shown in Figure 5.16.

2. To prepare air-free water for experiment (4), put distilled water into the flask, and boil gently for ten minutes. Add some nails, and boil again for five minutes. The nails must remain under the surface. Cool the flask quickly under the cold tap, and pour in a layer of liquid paraffin 1 cm deep. Close the flask with a rubber bung. Boiling drives out air, and the layer of oil keeps out air.

3. Allow the flasks to stand. You will be able to see some differences after two weeks, but if you can leave them for six months the results will be more interesting.

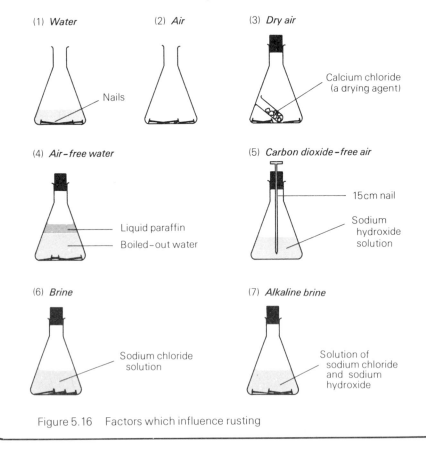

(1) *Water*   (2) *Air*   (3) *Dry air*

Nails

Calcium chloride (a drying agent)

(4) *Air-free water*   (5) *Carbon dioxide-free air*

Liquid paraffin
Boiled-out water

15cm nail

Sodium hydroxide solution

(6) *Brine*   (7) *Alkaline brine*

Sodium chloride solution

Solution of sodium chloride and sodium hydroxide

Figure 5.16   Factors which influence rusting

In Experiment 5.12, on rusting, you observed an increase in mass when iron rusts. You may have deduced that iron combines with some part of the water. There is air present as well as water, and the point of Experiment 5.13 was to find out whether air takes part in the reaction.

Your results should show that out of 100 cm³ of air, 21 cm³ are used up when iron rusts. This is the same figure as you obtained before for the percentage of oxygen in the air. The test with a lighted splint shows that there is no oxygen left for the splint to burn in. When iron rusts, it combines with oxygen in the air:

$$\text{Iron} + \text{Oxygen} \rightarrow \text{Iron oxide}$$

Since we live in a mechanical age, and since most of our machines are made of iron, rusting is a very important process, and we need to know as much as possible about it. Experiment 5.14 will have told you more. You will have found that: (1) in water, iron nails rust quickly; (2) in air, slow rusting occurs; and (3) in dry air there is no rust. There is no rust in (4), air-free water. The big nail in (5) has rusted above the cork but has not rusted in the carbon-dioxide-free air above the alkali. In (6), rusting is the most advanced: brine must encourage rusting. Number (7) shows that in alkali and brine together, the alkali wins, and rusting does not occur. The experiments prove that, in order to rust, iron needs air, including the acid gas carbon dioxide in the air, and water. Brine speeds up rusting, so that iron ships and iron bridges standing over seawater are very much at risk.

Methods of preventing iron from rusting

Since so many of the tools and machines we use are made of iron. or steel, which is an alloy of iron with other metals, it is important to think of ways of preventing iron from rusting. There are various methods.

Paint. Painting is the method used for large iron structures such as bridges and ships. The paint must be kept in good condition. If it chips and exposes the iron beneath, rusting will begin and will spread out underneath the remaining paint.

Oil or grease. A thin film of oil or grease will keep out air and water. This is suitable for moving parts of machinery, for example a bicycle chain.

Metal film. A thin film of another metal can protect iron from attack. Galvanised iron is iron which has been dipped into molten zinc and then lifted out. The layer of zinc protects the iron even if it becomes chipped. Galvanised iron is used for roofing and for dustbins. Food cans are made from iron coated with tin. If the tin coating becomes chipped, the iron rusts. Chromium plating is a process which is both protective and decorative and it is done by an electrical method. Car bumpers are treated in this way. Silver plating is a very expensive way of protecting iron and steel. It is done by electroplating and is used for decorative objects like vases and candlesticks, and for cutlery.

Stainless steel. Steels are alloys of iron containing some carbon with another metal. The choice of the other metal depends on what kind of steel is required. Some metals make a steel hard-wearing; some metals give a steel which can be sharpened to a knife edge. Stainless steel, which does not rust, is an alloy of iron, carbon and chromium. Air reacts with it to form a thin layer of chromium oxide, which protects the steel from any further attacks by air or water. Stainless steel is more expensive than iron and is only used when rusting would be disastrous. Cutlery is made of stainless steel.

Questions on Chapter 5

1. Explain the reasons for the following statements:

    (a) If air is bubbled through limewater for several minutes, the limewater turns milky.

    (b) When copper is heated in air, there is an increase in mass.

    (c) If iron nails are covered in a sealed tube with tap water they rust, but if the water is first boiled they do not rust.

    (d) A cold dish held in the yellow Bunsen flame soon becomes coated with a black substance.

2. Describe what you see happen when:

   (a) A strip of magnesium ribbon is set alight and then lowered into a jar of oxygen.

   (b) A piece of phosphorus in a dish, floating in a trough of water and covered with a bell jar, is touched with a hot wire, and the stopper of the bell jar is quickly replaced.

   (c) A combustion spoon is filled with sulphur, which is set alight and lowered into a jar of oxygen.

3. Name the three most abundant gases in the atmosphere and two gases which pollute the air.

4. Describe an experiment for finding the percentage by volume of oxygen in the air. What volume of nitrogen can you obtain from 250 cm^3 of air?

5.
Sulphur	Carbon	Oxygen
Sodium	Magnesium	Chlorine
Hydrogen	Nitrogen	Calcium

   From this list of elements, choose answers to the following questions:

   (a) Which two elements combine to form an edible substance?

   (b) Which elements are metals?

   (c) Which elements are non-metals?

   (d) Which two elements burn in air to produce oxides which dissolve in water to form acidic solutions?

   (e) Which element has to be kept out of contact with air?

   (f) Which two elements form an explosive mixture?

6. Supply words to fill the blanks in these sentences.

   'Calx' is the old name for _____.

   The discoverer of oxygen was _____.

   The first person to explain combustion was _____.

   The metal which combines with oxygen at 350 °C, and gives up oxygen at higher temperatures is _____.

7. You are given three gas jars. Describe three tests you would do to find out whether the gas they contain is oxygen.

8. When you look at your bicycle, you notice that different parts have received different treatments to prevent rusting. List these different treatments, and explain why different parts of the bicycle need to be treated in different ways.

Crossword on Chapter 5

Across

1   A greasy form of carbon (8)
3   Residue formed when some substances burn (3)
7, 21 across Poisonous gas from car exhausts (6, 8)
10 This metal is often coated with 5 down (4)
11 Animal living in sty (3)
12, 12 down This fraction of the air is 9 down (4, 6)
14 Symbol for indium (2)
15 A non-metallic element which burns with a yellow flame (10)
17 Metal used for container? (3)
19 Exists (2)
21 See 7 across
23 Devour (3)
25 Used to prevent rusting (5)
28 If you can't get oxygen you'll _____ breathing (4)
29 One of the ancient philosophers' four 'elements' (5)
31 Grains on the beach (4)
32 There is just a little of this gas in air (4)

Down

2. Another of the ancient philosophers' four 'elements' (3)
4   This element burns with a blue flame (7)
5   This metal is used for galvanising (4)
6   Used in flares, because of the way it burns (9)
8   An inert gas (5)
9   The 'lifeless' gas in the air (8)
12 See 12 across
13 Revolutions per minute (1, 1, 1)
16 Formed when 4 down and 6 down burn (6)
18 Priestley's great discovery (6)
20 It may be a stool in the lab (4)
22 The purest natural form of 29 across (4)
24 Weight (3)
26 North-west (1, 1)
27 29 across comes through this (3)
30 Royal Engineers (1, 1)

Trace this grid on to a piece of paper, and then fill in the answers.

6. Carbon

6.1 The forms of carbon

The element carbon exists in a number of different forms. They consist of carbon and nothing else, and yet they are different in appearance from one another. These forms of carbon differ in physical properties such as whether they are hard or soft, dull or shiny, but they have the same chemical reactions. The existence of forms of an element which are physically different but chemically identical is called *allotropy*. The two pure allotropes of carbon are *diamond* and *graphite*. There are many impure forms of carbon.

Diamond   The best specimens of diamond occur naturally. They are mined in South Africa and other places. Diamonds can be made artificially by subjecting graphite to high temperature and pressure; these artificial diamonds are not as large or as beautiful as natural diamonds. The great beauty of diamond, the sparkle and the glint of different colours of light, is due to the high refractive index of diamond, the ability to bend rays of light. Diamonds are very carefully cut under a microscope by trained craftsmen in order to get the angle between faces of the crystal which will give out the most light.

Diamonds which are not of high enough quality to make into jewellry and diamonds which have been manufactured have other uses. Diamond is the hardest naturally occurring material, and can be used to cut through softer materials. Diamonds are used for cutting glass and to tip the bits of drills which have to penetrate rock.

Graphite   Graphite is a dark grey, shiny material. When you touch it, a layer of graphite rubs off on to your fingers. Because of this, graphite is used as a lubricant, which shows how different it is from diamond, the hardest naturally occurring substance. As the quantity of graphite which occurs naturally is too small to satisfy the demand for it, it is manufactured by subjecting coke to high temperature and pressure.

Graphite conducts electricity. It is the only non-metallic element which does. It is used for the manufacture of electrodes to be used in batteries and cells. Another use of graphite is to mix it with clay and manufacture pencil 'leads', so called because lead also marks paper.

In addition to the two pure forms of carbon, diamond and graphite, there are many impure forms.

Impure forms

Wood charcoal is made by burning wood in a limited supply of air. This was traditionally done by covering a wood fire with turf to restrict the air supply. It is now done by burning wood in iron retorts, and Experiment 6.1 is a laboratory scale version of this method. Wood charcoal is used as a fuel, e.g. for barbecues, and as an artist's sketching material.

Animal charcoal is made by heating animal bones and remains in the absence of air. It is used as an *adsorbent*: the surface of the charcoal will attract a film of many substances, which are then said to be *adsorbed* (not *ab*sorbed) on the surface. It is used for adsorbing the colour from brown sugar to give us white sugar. Gas masks used in industry by workers dealing with poisonous gases contain charcoal to adsorb the gases. Experiment 6.2 works well with animal charcoal.

Lampblack is made by the partial burning of oil. It consists of very fine particles and is used for making printing ink and black paint. It is added to rubber to make tougher rubber tyres.

Soot is deposited on chimneys when fuels burn.

Gas carbon is left on the walls of retorts when coal is heated to form coal gas. It is used as a smokeless fuel.

Experiment 6.1

To prepare wood charcoal

1. Set up the apparatus shown in Figure 6.1, with the crucible lid restricting the flow of air to the wood shavings.

2. Heat strongly for ten minutes. Allow to cool.

3. Tip out the contents of the crucible. Collect the charcoal and use it in the next experiment.

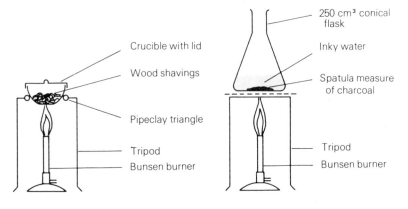

Figure 6.1   Preparation of wood charcoal   Figure 6.2   Inky water and charcoal

Experiment 6.2

To find out whether charcoal will adsorb the colour from inky water

1. Take 50 cm^3 of a mixture of ink and water ($\frac{1}{10}$ ink and $\frac{9}{10}$ water) in a 250 cm^3 conical flask. Add a spatula measure of charcoal.

2. Warm for ten minutes, as in Figure 6.2, and swirl the contents of the flask from time to time. Use the 'half-and-half' flame; there is no need to boil.

3. Let the contents of the flask cool. Filter.

4. Is the filtrate colourless? If it is, then the ink dye has been taken out of the solution by the charcoal, a process called adsorption.

6.2 Combustion of carbon

Experiment 6.3

To study the burning of charcoal

1. Take a piece of charcoal in tongs. Heat it strongly in a Bunsen flame until it glows red.

2. Take it out of the flame and blow on it vigorously. What do you notice? How can you explain this?

   If a substance is a fuel, once you get it hot, you can take away the source of heat and, provided you supply air, the substance will burn and give out heat. Would you say that charcoal is a fuel? If so, what makes you say that it is giving out heat? If you go on burning the piece of charcoal for some time, you will notice that it has become smaller. It makes you wonder what is formed when charcoal burns.

Experiment 6.4

To find out what is formed when charcoal burns

1. Set up the apparatus shown in Figure 6.3

2. Heat the charcoal until it glows red. Then remove the Bunsen, and keep the charcoal glowing red by pumping air over it with the hand-bellows.

3. Watch for any change in the limewater.

4. As a control experiment, pump air from the hand-bellows directly into the limewater for the same time as you ran the experiment.

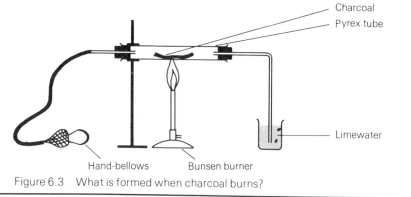

Figure 6.3   What is formed when charcoal burns?

An experiment on ourselves

1. Assemble conical flasks containing limewater as shown in Figure 6.4.

2. Six pupils line up in Row A, and six pupils line up in Row B.

3. At the word 'Go', Row A start sucking as fast as they can through the *short* tube as shown, and Row B start blowing through the *long* tube.

4. At the end of three minutes, compare the limewater in Row A with that in Row B.

5. Explain the difference.

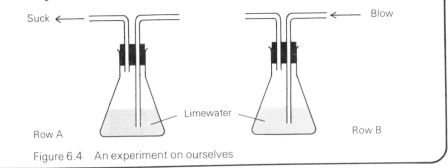

Figure 6.4   An experiment on ourselves

Experiments 6.3 and 6.4 will introduce you to this topic.

In Experiment 6.3, you found out that charcoal is a fuel. It burns and gives out heat, provided that it is at a high enough temperature and has a good supply of air. It has to be heated up first because no fuel will burn at a temperature below what is called its *ignition temperature*.

In Experiment 6.4, you burned charcoal, and saw limewater turn 'milky'; that is to say, white particles of solid appeared in it. Since no solid or liquid is entering the limewater, the change must be caused by a gas. The gas is formed when carbon burns. The active part of the air is oxygen, and carbon burns in oxygen to form a gas. This gas is carbon dioxide. The *di* in the name means two and tells you that one small particle of carbon called an atom has joined with two atoms of oxygen (see Chapter 7). When carbon dioxide reacts with limewater, which is a solution of calcium hydroxide, the white solid calcium carbonate is formed. Another name for it is chalk.

Carbon dioxide + Calcium hydroxide solution → Calcium carbonate
                         (Limewater)   (Chalk)

The test for carbon dioxide is that it turns limewater milky.

The control experiment showed the milkiness developing slowly when air was passed into limewater, proving that air contains carbon dioxide. Experiment 6.5 is about carbon dioxide in the air.

If you have done Experiment 6.5, you will have found that air sucked through limewater turns limewater slightly milky, but air which you breathe out turns limewater cloudy much faster. This shows that you must be breathing out air which contains much more carbon dioxide than ordinary air. Either you have been eating carbon or you have been eating foods which contain carbon and are oxidised to carbon dioxide in your body. You can find out by doing Experiment 6.6.

Figure 6.5

Preheater tower and kiln in cement works

An experiment to find out whether the foods we eat contain carbon

1. Take a piece of bread in tongs as shown in Figure 6.6 (a). Heat with a moderate flame until it has turned to a completely black mass and has stopped smoking. It is very important to heat until step (1) is complete.

2. Allow to cool, and put the black solid into an ignition tube.

3. Heat strongly until the black solid glows red. Squeeze the air out of a teat pipette, place it inside the ignition tube, and release the pressure on the teat to suck gas into the pipette. See Figure 6.6 (c).

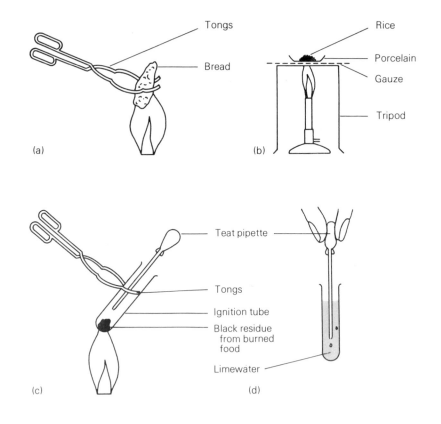

Figure 6.6   Experiment to find our whether foods contain carbon

4. Put some limewater into an ignition tube, and expel the gas from the teat pipette into the limewater, as shown in Figure 6.6 (d). Note whether it turns milky.

5. Repeat with another food. Macaroni and spaghetti can be held in tongs as in diagram (a). Rice and porridge oats can be heated on a piece of porcelain or ceramic paper as shown in Figure 6.6 (b).

6. If a food contains carbon, it will give carbon after heating by method (a) or (b), and this will be oxidised to carbon dioxide by steps (c) and (d). Make a list of the foods which you have found to contain carbon.

You have found out from this experiment that various foods you eat burn to form carbon dioxide. You will remember that carbon is a fuel: when it burns, energy is given out. You fuel your body with foods such as bread and porridge oats. These are oxidised by the oxygen you breathe in to give carbon dioxide, which you breathe out, and to provide you with energy. We talk about combustion of foods in the body rather than burning. Combustion is oxidation with or without a flame; in burning there is a flame.

All the foods, such as wheat, oats and rice, which you have experimented on are derived from plants. Plants build up these foods by the process of *photosynthesis*. *Photo* means *light* (from the Greek) and *synthesis* means *putting together*. The plant puts together carbon dioxide, taken in through the leaves, and water, taken in by the roots, and the energy of sunlight falling on the leaves. By means of the green substance in the plant called *chlorophyll*, which acts as a catalyst (see section 7.7), these three things are built up into sugar and oxygen. Sugar is converted to starch in the plant for storage. The process of photosynthesis can be represented as:

$$\text{Energy of sunlight} + \text{Carbon dioxide} + \text{Water} \xrightarrow[\text{in plants}]{\text{Chlorophyll}} \text{Sugar} + \text{Oxygen}$$

When animals eat foods containing sugar or starch, the energy stored in these foods, which came originally from the sun, is released by the process of respiration:

Oxygen + Sugar → Carbon dioxide + Water + Energy

Thus animals derive the energy they need from plant foods, and plants derive energy from the sun: plants *fix* the sun's energy, and animals *use* it. This energy cycle is shown in Figure 6.7.

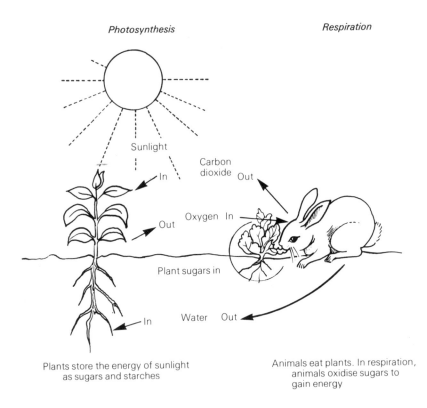

Photosynthesis   *Respiration*

Sunlight

Carbon dioxide Out

In

Oxygen In

Out

Plant sugars in

In   Water   Out

Plants store the energy of sunlight as sugars and starches

Animals eat plants. In respiration, animals oxidise sugars to gain energy

Figure 6.7   The energy cycle

You have seen how carbon takes oxygen from air to form carbon dioxide. In Chapter 5 you saw how metals combine with oxygen from the air. Let us now find out whether carbon has a greater or lesser liking for (in Chemistry called an *affinity* for) oxygen than metals have.

6.3 Reaction of carbon with metal oxides

Experiment 6.7

Will carbon take oxygen from metal oxides?

1. Assemble a number of metal oxides, zinc oxide, copper oxide, lead oxide, calcium oxide and others.

2. Take a charcoal block, and use an awl or a penknife to make a shallow hole in it.

3. Put in a little metal oxide. Dampen it with a drop of water to keep it from blowing away.

4. Light a Bunsen, and adjust it to the luminous flame; then let in a little air.

5. Taking a blowpipe, direct a flame on to the metal oxide in the carbon block. Pause for breath, and continue (see Figure 6.8).

6. If you have bushy hair, you may well singe it in this experiment. Pin it back!

7. Allow the carbon blocks to cool for some hours before putting them away.

8. Tabulate your observations under the headings:
   Metal oxide   Appearance   Observations

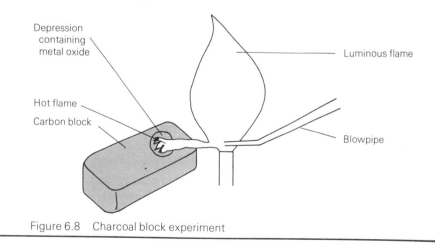

Figure 6.8   Charcoal block experiment

You will have found out from Experiment 6.7 that carbon takes oxygen away from some metals, such as copper and lead, but is unable to take oxygen away from others such as zinc and calcium:

Carbon + Copper oxide → Copper + Carbon dioxide

Carbon + Lead oxide → Lead + Carbon dioxide

The process of taking oxygen away is called *reduction*; it is the opposite of giving oxygen, which is called *oxidation*. Carbon reduces lead oxide to lead. Lead oxide oxidises carbon to carbon dioxide. The processes of oxidation and reduction always occur together. *A substance which gives oxygen is called an oxidising agent; a substance which takes oxygen is called a reducing agent.* Lead oxide acts as an oxidising agent, carbon acts as a reducing agent; in the reaction the oxidising agent is reduced, and the reducing agent is oxidised.

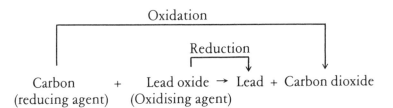

Oxidation

Reduction

Carbon + Lead oxide → Lead + Carbon dioxide
(reducing agent) (Oxidising agent)

It is possible to put elements in order of their affinity for (liking for) oxygen. When you do this, you find that some metals are above carbon, and some are below carbon. Carbon will reduce the oxides of metals which are below it in order of affinity for oxygen.

6.4 Carbonates

A carbonate is a compound which contains a metal and carbon and oxygen. Whenever you see a name ending in *-ate*, it means that the compound contains oxygen. Experiment 6.8 involves heating a number of carbonates to find out which decompose to give carbon dioxide.

Experiment 6.9 is a more detailed study of the effect of heat on calcium carbonate. Calcium carbonate is the most widespread of metal carbonates. It is found in three forms, marble, limestone and chalk. They look different, but are chemically the same.

Action of heat on carbonates

1. The method is shown in Figure 6.9 (a) and (b). Put a little copper carbonate into the ignition tube in (a), hold the tube in tongs and heat.

2. Test for carbon dioxide by using a teat pipette to suck in gas in (a) and expel it into limewater in (b).

3. Repeat, using other metal carbonates, and tabulate your results.

Table 6.1   *Action of heat on carbonates*

Carbonate	Appearance	Is carbon dioxide evolved on heating?	Observations

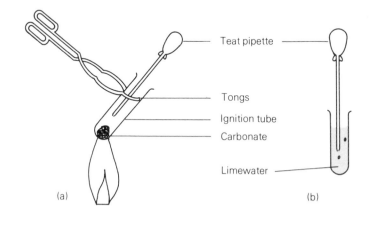

(a)   (b)

Figure 6.9   Heating carbonates

Experiment 6.9

Action of heat on calcium carbonate

1. Assemble the apparatus as shown in Figure 6.10. Put the piece of marble on the edge of the gauze, where it can be heated strongly. Balance it carefully: you do not want it to fall off when hot.

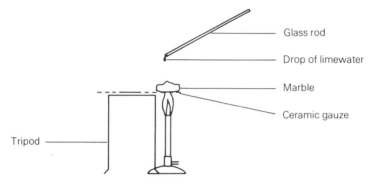

Figure 6.10   Heating calcium carbonate

2. Heat the marble with a roaring Bunsen flame for fifteen minutes. From time to time, hold a drop of limewater on the end of a glass rod over the marble. Observe the limewater.

3. Allow to cool. When cool, put the piece of marble into an evaporating basin. Add one drop of universal indicator. Compare the colour with a drop of universal indicator on an unheated piece of marble. What does the colour tell you?

4. Carefully, from a teat pipette, add water drop by drop. You should notice two things: (a) a gas, (b) a change on the surface of the marble. What makes you think that there is a change in energy occurring?

5. Dissolve the crumbly solid formed on the surface of the marble in water. Filter. Test the solution.

   (a) To a portion, add universal indicator.

   (b) Take a second portion, and blow into it through a straw. Do you recognise this solution?

6. Explain your observations. What gas was evolved? What formed at the surface of the marble? What happened when you added water? What solution was formed?

In Experiment 6.8, you will have found that many metal carbonates give carbon dioxide on heating. Some need to be heated for longer than others to make them decompose.

When calcium carbonate is heated strongly, as in Experiment 6.9, it splits up to give carbon dioxide and calcium oxide, a white solid which turns universal indicator blue, the alkaline colour. When water is added to calcium oxide, you see it turn to steam. This is because the reaction between calcium oxide and water gives out heat; it is an exothermic reaction. In the process, calcium hydroxide is formed. When you make a solution of calcium hydroxide and blow into it, it turns milky, enabling you to recognise it as limewater. The reactions are:

Calcium carbonate → Calcium oxide + Carbon dioxide

Calcium oxide + Water → Calcium hydroxide + Heat

Calcium hydroxide + Water → Calcium hydroxide solution
(Limewater)

It is worth repeating the addition of water to calcium oxide on a larger scale because in this experiment calcium oxide forms only on the surface of the marble.

If you do, make sure the calcium oxide is dry by warming it gently for a few minutes. Then allow it to cool. Add water dropwise from a teat pipette, and watch the vigorous reaction which occurs.

Early chemists, such as Joseph Black who did a lot of work on calcium carbonate around 1755, thought that while calcium oxide was absorbing water and swelling and giving out heat, it looked as though it were alive. They called it *quicklime*, *quick* being a mediæval word for *living*. Being so eager to combine with water, quicklime appears to be *thirsty*. When it has satisfied its desire for water, it appears to have *slaked* its thirst. Calcium hydroxide is therefore called *slaked lime*. A solution of slaked lime is called *limewater*.

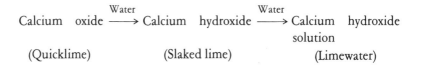

93

Manufacture of quicklime

Quicklime is made by heating limestone in a furnace at 1000 °C, called a lime kiln. Limestone and coke are fed in at the top. A good draught of air comes in at the bottom to burn the coke. Carbon dioxide produced when limestone decomposes is carried away at the top by the stream of air. Quicklime is taken out at the bottom (see Figure 6.11).

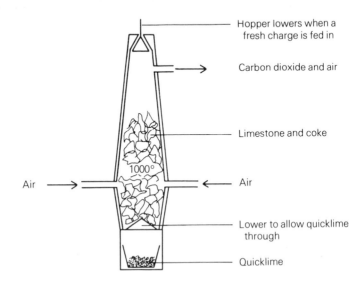

Figure 6.11   A lime kiln

Quicklime is used in vast quantities on building sites. It is slaked and then mixed with sand to form mortar. Mortar holds bricks together. It gradually hardens on exposure to air, as carbon dioxide in the air turns slaked lime to calcium carbonate. Other materials used on building sites are cement, which is made from crushed limestone and water and clay, and concrete, which is cement to which sand and stones have been added.

Figure 6.12 shows a view of the Blue Circle Group Hope Works, where cement is made. Limestone is quarried in the hills behind the works. At the extreme right of the works is a very tall chimney stack. This is in the preheater building, and the kilns are the two horizontal pipes which connect the building and the rest of the plant. Figure 6.5 shows a close-up of the preheater tower and one of the kilns.

We have seen how it is possible to obtain carbon dioxide by heating limestone or marble to 800–1000 °C for some time. A more convenient method is to add a dilute acid. If you add marble chips to the three most common acids, you will find that dilute hydrochloric acid and dilute nitric acid give carbon dioxide quickly, without heating. Dilute sulphuric acid reacts for a few seconds, and then slows down and stops. This is because calcium carbonate reacts to form calcium sulphate. Calcium sulphate is insoluble, and the insoluble film of calcium sulphate round the marble chip protects it from further attack by the sulphuric acid.

You can now prepare carbon dioxide by the action of dilute hydrochloric acid or dilute nitric acid on marble chips:

| Calcium | + | Hydrochloric | → | Calcium | + | Carbon |
| carbonate | | acid | | chloride | | dioxide |

Calcium carbonate + Nitric acid → Calcium nitrate + Carbon dioxide

Figure 6.12   Blue Circle Group Works at Hope

6.5 Carbon dioxide

To prepare carbon dioxide and study its properties

1. Set up the apparatus as shown in Figure 6.13.

2. Add dilute hydrochloric acid to the marble chips. Discard the first boiling tube of gas collected as it will contain a lot of air.

3. Collect two boiling tubes of gas over water as shown in method (a). Cork the tubes before taking them out of the trough. Stand them in a rack.

4. Attach a right-angled delivery tube as in method (b). Collect two gas jars full of gas by downward delivery. A slow count of fifteen should tell you when all the air has been driven out. Cover the gas jars with lids.

5. Leave the right-angled tube attached, and bubble the gas through limewater in a boiling tube.

6. Bubble the gas through a boiling tube one third full of water to which universal indicator has been added.

7. Into a boiling tube of gas, lower a burning taper.

8. Lower a burning strip of magnesium, held in tongs, into a boiling tube of the gas. To the solid formed in the boiling tube, add dilute hydrochloric acid. Filter. What remains in the filter paper? Can you name this substance?

9. Figure 6.14 (a) shows how to find out whether carbon dioxide is denser or less dense than air. Put a jar of carbon dioxide over one of air. Remove the lids, count twenty seconds, and replace the lids. Test both jars with a lighted splint. Where is the carbon dioxide?

10. Repeat (9) with carbon dioxide at the bottom and air at the top, as in Figure 6.14 (b). What happens to the carbon dioxide?

This test can be done with stoppered boiling tubes if there are not enough gas jars for the class.

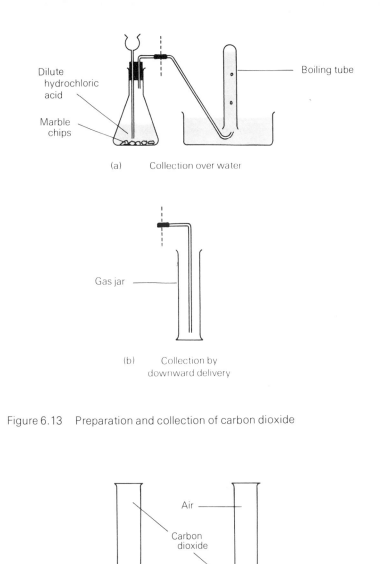

Figure 6.13   Preparation and collection of carbon dioxide

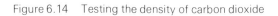

Figure 6.14   Testing the density of carbon dioxide

You have found that carbon dioxide is a colourless, odourless gas which is not very soluble in water: it can be collected over water. It is an acid gas. When bubbled through universal indicator solution for some time, it turned the indicator a cherry red, showing that a weakly acidic solution had been formed.

Carbon dioxide does not support the combustion of a burning taper in (7). In (8), however, burning magnesium continued to burn when lowered into carbon dioxide. A white solid and a black solid were formed. The white solid dissolved in dilute hydrochloric acid, and the black solid did not. Starting from magnesium and carbon dioxide, we can obtain a white solid which reacts with acid and a black solid if we suggest:

Magnesium + Carbon dioxide → Magnesium oxide + Carbon

Magnesium burns with a flame which is hot enough to split up carbon dioxide into carbon and oxygen. It burns in the oxygen to form magnesium oxide, and leaves carbon behind. Magnesium is one of the elements which has a greater affinity for oxygen than carbon has.

Test (9) shows that carbon dioxide sinks into the bottom gas jar, and test (10) shows that, when placed at the bottom, carbon dioxide will stay in the bottom gas jar. Carbon dioxide is denser than air.

6.6 The carbon dioxide cycle

Carbon dioxide makes up 0.03% of the air, a very small but essential percentage. The level of carbon dioxide is kept constant by a balance between processes which take carbon dioxide from the air and processes which put carbon dioxide into the air. We and other animals breathe out carbon dioxide into the air. Plants take carbon dioxide from the air to use in photosynthesis. Combustion of coal and oil produces carbon dioxide. Decay of plant and animal matter produces carbon dioxide. Carbon dioxide dissolves in the sea, and is built up into the shells of sea creatures as calcium carbonate. When these creatures die, they sink to the sea bed. There they accumulate to form deposits of limestone. Limestone is heated in kilns to give carbon dioxide. All these processes can be represented in a diagram, such as the one in Figure 6.15. The balance between reactions using carbon dioxide and reactions producing carbon dioxide is called the carbon dioxide cycle.

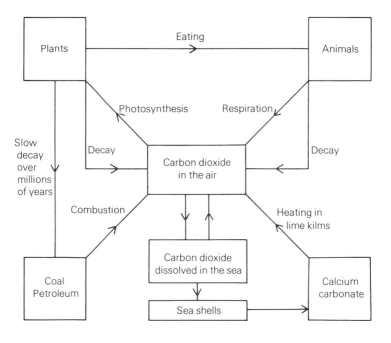

Figure 6.15 The carbon dioxide cycle

6.7 Methods of fighting fires

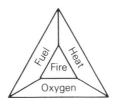

Figure 6.16 The fire triangle

The fire triangle shows the three things a fire needs: fuel to burn, heat to raise the fuel to the temperature at which it is hot enough to burn (called its *ignition temperature*), and oxygen. If any one of these sides of the triangle is removed, the fire will go out.

Removal of fuel

A fire will burn itself out when it runs out of fuel. This technique is used in fighting forest fires. If a wide swathe of trees and bushes in the path of the fire is removed, the fire will not be able to spread and will die out.

Removal of heat

Throwing water onto a fire cools it down. If it cools down below the ignition temperature of the fuel, the fire will go out. There are some chemical fires on which water cannot be used, for example sodium fires. Water reacts with sodium to give the flammable gas hydrogen, and this makes the fire worse. It is part of a fireman's training to learn which chemicals react with water.

Electrical fires cannot be extinguished with water. If you direct a jet of water on to a smouldering piece of electrical equipment, a current of electricity can pass along the jet of water and give you a severe shock. The method to use here is to switch off at the mains and use one of the carbon dioxide fire extinguishers.

Oil fires cannot be extinguished with water. This is why it is so dangerous if the oil tanks on a ship catch fire. The burning oil floats on top of the water, and the blaze spreads over a wide area. If the crew try to abandon the ship, they face the danger of being burned in their lifeboats.

Experiment 6.11 shows what happens when you try to extinguish an oil fire with water, and also how you can put out the fire with a damp cloth. The fire goes out because the cloth is excluding air. The cloth is dampened to prevent it from catching fire itself. This method is the one to use if a pan of cooking fat catches fire at home: switch off the gas or electricity and cover the pan with a damp towel.

For many fires, however, water is a good extinguisher. The soda–acid type of extinguisher provides a means of obtaining a powerful jet of water quickly. This type of extinguisher is the large kind which stands on the floor. To operate it, you have to drive in a knob. This breaks a vial of concentrated sulphuric acid in a container filled with a solution of sodium hydrogencarbonate (called soda for short). Carbon dioxide is formed instantly, and the pressure of the gas forces a jet of the solution out of a nozzle. Figure 6.17 shows a commercial soda–acid extinguisher, and Figure 6.18 shows a model which you can set up.

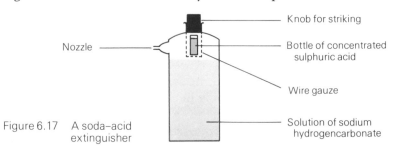

Figure 6.17   A soda–acid extinguisher

Knob for striking

Nozzle

Bottle of concentrated sulphuric acid

Wire gauze

Solution of sodium hydrogencarbonate

Experiment 6.11

Will water extinguish oil fires?

1. Put an evaporating basin on to a tile on the bench. Put into it 20 cm³ of oil (cooking oil or turpentine) and, using a piece of rag as a wick, set fire to it.

2. Now direct a jet of water from a squeeze bottle on to the flames.

3. Do you find that the oil floats on top of the water and continues to burn?

4. Take a bench cloth, wet it under the tap, wring it out and put it over the evaporating basin.

Experiment 6.12

To try out a model of a soda–acid fire extinguisher

1. Fill a conical flask two thirds full of saturated sodium hydrogen-carbonate solution. Fit a delivery tube through a rubber bung to reach down to near the bottom of the flask (see Figure 6.18).

2. Tie a piece of cotton thread round the mouth of an ignition tube. Wearing safety glasses, fill the ignition tube with concentrated hydrochloric acid, and lower it into the conical flask. Keep the acid above the level of the solution so that they do not mix. Replace the bung.

3. Now you can try out your model. Take a small piece of cotton wool in an evaporating basin. Add 10 cm³ of methylated spirits and set fire to it.

4. Now shake the model fire extinguisher so that the acid mixes with and then reacts with the solution of sodium hydrogencarbonate. Direct the jet of water on to the fire.

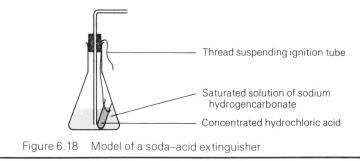

Thread suspending ignition tube

Saturated solution of sodium hydrogencarbonate

Concentrated hydrochloric acid

Figure 6.18   Model of a soda–acid extinguisher

Removal of oxygen

The third side of the triangle is oxygen. Carbon dioxide extinguishers act by excluding oxygen. Carbon dioxide is a dense gas which does not support combustion. It forms a blanket over the fire and keeps out oxygen. Carbon dioxide is a good extinguisher for many types of fire, including oil fires and electrical fires. A fire of burning magnesium, however, would continue to burn in the gas. There are three types of extinguisher using carbon dioxide. These are *powder, gas* and *foam extinguishers.*

Powder extinguishers contain substances like sodium hydrogen-carbonate (baking soda), which decomposes in the heat of the fire to give carbon dioxide. This drives away the air, and the fire goes out. Powder extinguishers are useful for electrical fires. They will not cope with a big fire. They are small in size and usually placed in a bracket on the wall. Experiment 6.13 uses a powder extinguisher.

Cylinders of carbon dioxide stored under pressure are useful fire extinguishers. Figure 6.19 shows an extinguisher of this type. It is usually operated by removing a pin and squeezing the trigger. A jet of carbon dioxide gas comes out of the nozzle for about two minutes. One must never use a fire extinguisher except when it is needed for putting out a fire. If there is a cylinder of carbon dioxide in your store, your teacher may demonstrate how it works on some burning kerosene.

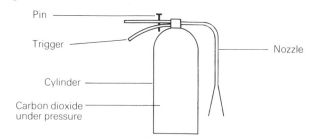

Figure 6.19   A carbon dioxide fire extinguisher

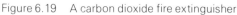

Experiment 6.13

To use a model powder extinguisher

1. Get a little kerosene (called 'paraffin' by retailers) burning, with a wick, in an evaporating basin.

2. With a spatula, sprinkle sodium hydrogencarbonate onto it.

3. Is the fire extinguished?

Experiment 6.14

To use a model foam extinguisher

1. Assemble the model shown in Figure 6.20. The bottle contains 100 cm³ saturated sodium hydrogencarbonate solution containing 1 g saponin, a foam stabiliser. The test tube contains 10 cm³ of dilute hydrochloric acid.

2. Start a small fire in an evaporating basin, using 10 cm³ kerosene (paraffin) and a wick (a piece of rag or filter paper).

3. Turn the model foam extinguisher upside down, and direct the stream of foam on to the fire.

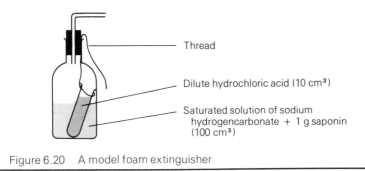

Thread

Dilute hydrochloric acid (10 cm³)

Saturated solution of sodium hydrogencarbonate + 1 g saponin (100 cm³)

Figure 6.20   A model foam extinguisher

Experiment 6.15

To use carbon dioxide gas as a fire extinguisher

1. Assemble the equipment shown in Figure 6.21. Have to hand a bottle of dilute hydrochloric acid.

2. Set fire to the kerosene.

3. Pour acid carefully down the side of the big beaker on to the marble chips.

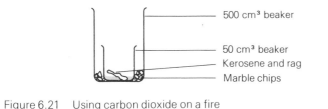

500 cm³ beaker

50 cm³ beaker
Kerosene and rag
Marble chips

Figure 6.21   Using carbon dioxide on a fire

Foam extinguishers employ carbon dioxide. The principle is that carbon dioxide bubbles mixed with foam form a heavier layer on top of the fire than carbon dioxide alone. This type of extinguisher is large and usually stands on the floor. When needed, it is inverted and tapped sharply on the floor to push in a knob. The knob releases an acid solution into a solution of sodium hydrogencarbonate and *saponin*, a foam stabiliser. A jet of foam consisting of the solution and carbon dioxide is expelled. Experiment 6.14 shows you how to make and use a model foam extinguisher.

A foam extinguisher is unsuitable for those types of chemical fires which react with water.

Experiment 6.15 is a method of using carbon dioxide to put out a fire. You will see effervescence as the acid reacts with marble to form carbon dioxide. Being denser than air, carbon dioxide sinks into the small beaker and displaces air. The fire goes out for lack of air.

There are gases other than carbon dioxide which do not support combustion and which are denser than air. Pyrene extinguishers contain *tetrachloromethane* (often called *carbon tetrachloride*). On hot surfaces, this liquid produces a dense vapour which excludes air from the fire. On extremely hot surfaces, it can be oxidised to the poisonous gas carbonyl chloride (often called phosgene). An improvement is the use of BCD extinguishers. The liquid which they contain, bromochlorodifluoromethane, forms a dense vapour which is chemically unreactive even at high temperatures.

6.8 Fuels

Coal

Many of the fuels we use contain carbon and carbon compounds. Coal is one. You have all seen coal burning with a good supply of air. The next experiment shows what happens when coal is heated in the absence of air.

Experiment 6.16

The action of heat on coal in the absence of air

1. Set up the apparatus as shown in Figure 6.22. Heat gently, and make a note of anything you see.

2. If you see any gas coming out of the side-arm, try to light it with a splint. Note whether the gas burns and, if it does, note the appearance of the flame.

3. Finally, heat very strongly for a few minutes.

4. Allow the apparatus to cool, then remove the side-arm tube. Make a note of what you see in the water. Test the water with a piece of neutral litmus paper. There are two products for you to detect in the side-arm tube.

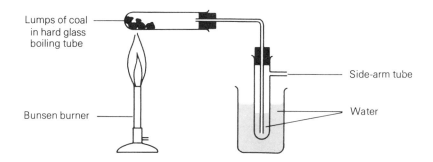

Figure 6.22   Heating coal in the absence of air

5. When the boiling tube is cool enough, tap out the residue. Make a note of its appearance. What do you think it is?

6. Put a little coal on a tin lid. Heat strongly, and make a note of how well it burns and what kind of flame it has.

7. Put a little of your residue on a tin lid. Heat strongly, and notice whether it burns well and with what sort of flame.

8. Say which you think would make the better fuel – coal or the residue – and give your reasons.

9. Would you say that the coal in the boiling tube had been burned?

10. Name four products formed by the action of heat on coal in a very limited supply of air.

11. You can repeat the experiment with wood instead of coal.

Coal is heated industrially in iron retorts in order to make coal gas. At one time, industrial and domestic users of gas relied heavily on coal gas, but recently, in practically the whole of the country, North Sea gas has replaced coal gas. In addition to coal gas, coke is produced: this is the residue in the retorts (and in your boiling tube). It is a useful smokeless fuel. Since coke is coal from which coal gas and tar have been removed, it does not give out as much heat as coal on burning. Coal-tar is a third product. In your experiment, you see it as a thick black layer on top of the water in the side-arm tube. By distillation, it can be separated into a number of substances: solvents such as benzene and toluene, antiseptics such as phenol, creosote for protecting wood, pitch for tarring roads. Ammoniacal liquor is a further product. You may have found that the water in the side-arm tube became alkaline. This is because the alkaline gas ammonia is given off when coal is heated. Ammoniacal liquor is used in the manufacture of household ammonia for cleaning and in the manufacture of fertilisers.

For an interesting account of the formation of coal, methods of mining, the coal gas industry and how electricity is generated by burning coal, read the Nuffield Foundation Chemistry Background Book, *Coal*.

Petroleum

Petroleum oil is a mixture of a large number of compounds. Most of them are compounds of carbon and hydrogen, called hydrocarbons. Crude oil is separated into a number of different parts or fractions by distillation. The crude oil is vaporised and then fed into a high cylindrical column. Some compounds, with low boiling points, find it easier to remain gases than compounds with high boiling points, and they therefore pass up to the top of the column while the condensed higher boiling point compounds are trickling down. Perforated plates all the way up the column ensure good contact between the gases going up and the liquids trickling down. The result is that fractions of different boiling points can be tapped off from different heights of the column. Each fraction is not a pure compound but a mixture of compounds of similar boiling points. Figure 6.23 shows a diagram of a fractionating column in an oil refinery and the names of the fractions collected.

Petroleum gases are sold as bottled gases, such as Camping Gaz. There is a huge demand for the next five fractions, aviation fuel, gasoline for motor vehicles, kerosene for jet engines and rockets, and gas oil for heavy vehicles like tractors and diesel oil for trains.

The high boiling point fractions are used as lubricating oil and as wax for candles and waxed paper. The liquid in the bottom of the column is used as very heavy fuel oil in ships and as bitumen for use on the roads. All these fractions have to be purified before use to remove ingredients which could cause pollution when the fuel oils are burned.

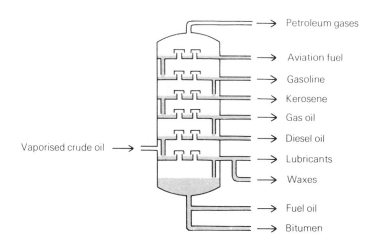

Figure 6.23   A fractionating column in an oil refinery

For an account of the formation of petroleum oil, methods of prospecting and drilling for oil and the petrochemicals industry, see the Nuffield Foundation Chemistry Background Book, *Petroleum*.

In the next experiment, we use a candle made from paraffin wax, one of the high boiling point fractions obtained from the distillation of crude oil, in order to find out what is formed when petroleum fractions burn.

To find out what is formed when a candle burns

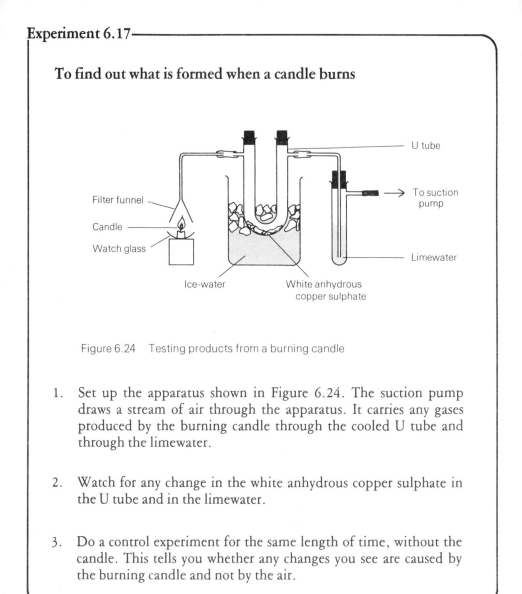

Figure 6.24   Testing products from a burning candle

1. Set up the apparatus shown in Figure 6.24. The suction pump draws a stream of air through the apparatus. It carries any gases produced by the burning candle through the cooled U tube and through the limewater.

2. Watch for any change in the white anhydrous copper sulphate in the U tube and in the limewater.

3. Do a control experiment for the same length of time, without the candle. This tells you whether any changes you see are caused by the burning candle and not by the air.

The combustion of petroleum fractions, such as gasoline and kerosene, to give out energy is a very important reaction. Our cars and our planes and some of our central heating systems depend on it. With so much hydrocarbon material being burned, it is important to know what is formed when hydrocarbons burn. Experiment 6.17 used a candle made from paraffin wax, one of the materials obtained from the distillation of crude oil.

If you found that air drawn over the burning candle turned anhydrous copper sulphate blue and limewater milky, you detected water and carbon dioxide. Your control experiment showed that these gases are present in the air. They are detected much faster when you have the candle burning: candle wax burns to form carbon dioxide and water. All petroleum fractions are hydrocarbons, compounds of carbon and hydrogen, and all burn to form carbon dioxide and water.

If the supply of air is limited when petrol and other hydrocarbon fuels burn, combustion may not be complete. Instead of burning to form carbon dioxide, which is produced when combustion is complete, they may burn to form carbon and carbon monoxide. You will have noticed some soot, which is a form of carbon, in the candle experiment. Carbon monoxide is a poisonous gas. Its name means that every tiny particle or atom of carbon in it has combined with one tiny particle or atom of oxygen. In carbon dioxide, each atom of carbon has combined with two atoms of oxygen. Atoms are discussed in Chapter 7. Running a car in a closed garage does not provide the engine with enough air to burn the fuel completely to carbon dioxide and water. Instead it burns incompletely to carbon monoxide and water, and the exhaust fumes will contain carbon monoxide. Soon there will be a dangerous level of this poisonous gas.

If fuel oils contain sulphur compounds, the poisonous gas sulphur dioxide will be formed when the oils are burned. The oil burned in domestic heaters must be carefully freed from sulphur compounds. It is called *kerosene* in the oil industry and *paraffin* by the retailers who sell it. Other fuel oils are allowed to have a higher sulphur content, and in busy cities with traffic congestion, there can be an unpleasant build-up of sulphur dioxide. To avoid this pollution, the oil industry will have to do a more careful and therefore a more expensive purification of the fuel oils. The industry can do this if people are prepared to pay more for petrol.

Natural gas Natural gas is the name given to the gas which always occurs together with crude petroleum oil. Since it was discovered in the North Sea, natural gas has been piped into Britain, and it is now extensively used for cooking and heating. Natural gas is mainly methane, a compound of carbon and hydrogen.

Questions on Chapter 6

1. Write down a word or words to fill each blank:

   The hardest naturally occurring substance is _____. It is a form of the element _____. It is used in jewellery because _____, and it is used for industrial purposes such as _____. The other pure form of this element is _____. When baked with clay, this is used to make _____. It conducts electricity, and is therefore used for _____. Since it is soft and flaky, it is used as _____.

2. Write down a word or words which will correctly fill the blanks in this passage:

   Some foods that we eat contain starch. Examples are _____, _____ and _____. When these foods are burned in the laboratory a black mass of _____ is left. This can be oxidised on heating to form the gas _____, which turns limewater _____. In our bodies, the starchy foods we eat are oxidised by _____ in the air we breathe in. This is why we breathe out _____. The energy stored in starch is put there by the process of _____, in which plants use _____ and water, and the energy of sunlight. This process is catalysed (helped) by _____ in the plants.

3. Name two allotropes of carbon. How do they differ? Give two uses for each allotrope. Name two impure forms of carbon. Describe how they are made. Give a use for each of them.

4. Describe what you see:

   (a) when carbon is heated;

   (b) when a mixture of carbon and lead oxide is heated;

   (c) when magnesium is burned in carbon dioxide;

   (d) when water is added to cooled, freshly made lime.

   Explain what chemical reactions take place.

5. Give the chemical names for limestone, slaked lime, quicklime and limewater.

6. The chemicals used in the laboratory preparation of carbon dioxide are calcium carbonate and dilute hydrochloric acid. Explain why dilute sulphuric acid is not used. Explain why carbon dioxide is collected over water and why it can be collected by downward delivery. Draw an apparatus you could use for the preparation and collection of carbon dioxide.

7. (a) A green powder when heated gives off a colourless gas that turns limewater milky. Name the solid, and name the gas.

    (b) A black powder will decolourise litmus solution. Name this solid.

    (c) A white solid becomes hot when water is added. Name this solid.

    (d) Two black solids, when heated together, form a reddish–brown solid and the gas which turns limewater milky. Suggest what these two black solids might be.

8. Mortar consists of a wet mixture of calcium hydroxide, sand, and animal hair. If some old mortar is powdered and treated with dilute acid, carbon dioxide is formed. Where do you think the carbon dioxide *originally* came from?

9. (a) What are the two liquids in a soda–acid extinguisher? Which gas do they react to form?

    (b) How does water extinguish a fire? Explain why water cannot be used to extinguish a pan of burning oil.

    (c) How would you extinguish a chip pan fire if no extinguisher were available?

    (d) Name two types of extinguisher which are safe to use on burning oil.

    (e) Give an example of another kind of fire on which water cannot be used. Explain why.

    (f) Make a list of ten simple precautions that can be taken in the home to prevent fire.

10. Write down the missing words:

Adding oxygen to a substance is called _____.

Taking oxygen from a substance is called _____.

When carbon reacts with copper oxide to form copper, carbon is acting as a _____ agent. Carbon is _____ to carbon dioxide, and copper oxide is _____ to copper.

11. You are supplied with gas jars containing carbon dioxide. Describe experiments you can do to show that:

(a) carbon dioxide is denser than air;

(b) carbon dioxide dissolves in sodium hydroxide solution;

(c) carbon dioxide will not support combustion.

12. Name eight substances which are obtained from petroleum oil, and give their uses.

13. What are the two chief substances produced when petrol burns? What other substance is produced if the supply of air is restricted? What pollutant is formed in small amounts when petrol burns?

Crossword on Chapter 6

Carbon and Carbon Dioxide

Across

1 Residue obtained by heating coal (4)
2 They are used to cover gas jars (4)
5 Tool sometimes tipped with 6 across (5)
6 The hardest naturally occurring material (7)
8 A process which resembles burning, without a flame (10)
10 Given out in 8 across (4)
11 Only about 20% of it is oxygen (3)
12 It is given out when magnesium burns (5)
13, 6 across will _____ 12 across (7)
17 Sweet substances containing carbon (6)
21 These burn to give carbon dioxide and 19 down (13)
24 Carbon _____ copper oxide to copper (7)
26 Sodium hydroxide is caustic _____ (4)
27 You obtain this from foods containing 21 across (6)

Down

1 Form of carbon used by artists (8)
2 A name for calcium carbonate (9)
3 Slippery form of carbon (8)
4, 3 down and 6 across are a pair of these (10)
7 Carbon _____ is formed when carbon burns (7)
9 Beautiful form of calcium carbonate (6)
14 Name given to a mixture of clay and 3 down (4)
15 Symbol for caesium (2)
16 The French for *you* (2)
18, 3 down feels like this (6)
19 Formed by 8 across of 21 across (5)
20 It gives you 12 across (and some 10 across) (5)
22 A jelly-like substance, used in biology laboratories (4)
23 This fuel is burned in moorland areas of Ireland (4)
25 South-east (1, 1)

Trace this grid on to a piece of paper, and then fill in the answers.

Trace this grid on a piece of paper, and then fill in the answers.

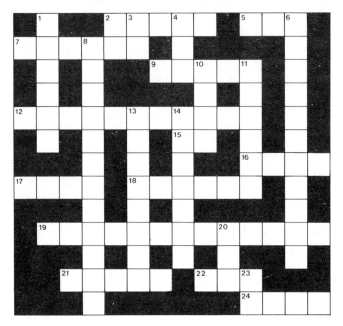

Crossword on Chapter 6

Fuels

Across

2, 20 down, 5 across Source of petrol in Britain (5, 3, 3)
5 See 2 across, 6 down and 11 down
7 Formed when substances burn (6)
9 It is used to support apparatus (5)
12, 8 down Used on coal to give valuable products (11, 12)
15 Short form of education (2)
16 An important school test (4)
17 See 1 down
18 Obtained by 8 down of 2 across, 20 down, 5 across (6)
19 Putting out fire (like carbon dioxide does) (13)
21 Use 22 across to cover these (5)
22 Left behind after 8 down is used on 2 across, 20 down, 5 across (3)
24 See 1 down

Down

1 17 across, 24 across Sides of fire triangle (6, 4, 4)
2 Opposite of south-west (1, 1)
3 Outsize (abbreviation) (2)
4 Powerful explosive (1, 1, 1)
6, 5 across Use this to get a smooth-running engine (11, 3)
8 Method of separating immiscible liquids (12)
10 The sort of chemical that liberates carbon dioxide from carbonates (4)
11, 5 across Trains use _____ _____ (6, 3)
13 Still closed (8)
14 A preliminary trial (4, 3)
20 See 2 across
23 Short for rhesus factor (2)

7. Atoms and molecules

7.1 Dalton's Atomic Theory

For many centuries, there were two different theories about the nature of matter. Some people believed that matter is *continuous*, that it is the same all through without any spaces. Others held that matter is *particulate* – that is, made up of tiny pieces or particles with spaces in between them. Democritus, in 500 BC, was the first to suggest that matter is composed of tiny particles. The idea could not be put to the test, and was not accepted. Hundreds of years went by before it was revived in 1808 by a British chemist called Dalton. He was able to show how the idea that matter is composed of minute particles made sense of a large body of chemical knowledge.

Dalton called the particles *atoms*, a Greek word meaning *cannot be divided*. Dalton's Atomic Theory can be stated in three parts:

(1) Matter is composed of a large number of atoms, which are minute particles which cannot be created or destroyed or divided.
(2) All the atoms of an element are identical in every way. The atoms of an element differ from the atoms of all other elements.
(3) Chemical combination takes place between small whole numbers of atoms of the elements concerned. For example, one atom of oxygen combines with two atoms of hydrogen. Atoms combine to form molecules. All the molecules of a compound are alike in every way.

You will see that in part (1) Dalton says that atoms cannot be created or destroyed. The reason he suggested this was that in a large number of chemical reactions, the starting materials, called reactants, had been weighed and the products had been weighed, and it had always been found that the mass of the reactants was equal to the mass of the products. If matter is composed of atoms which cannot be created or destroyed, then a chemical reaction involves a rearrangement of atoms, some chemical bonds between atoms being broken and new bonds formed, without any loss or gain in mass.

115

It follows from part (1) of Dalton's theory that the matter of which our world is composed is the same as that in existence at the beginning of time. The oxygen in the air you are breathing in at this moment may have been breathed in by King John as he signed the Magna Carta in 1215 or by King Harold as he fought the invading Normans in 1066. Through a series of chemical reactions, the oxygen they breathed in became free oxygen once more and returned to the atmosphere ready for you to breathe in.

It may be useful to list the meanings of some of the words Dalton used.

Element. An element is a substance which cannot be broken down into simpler substances by any known chemical means. It consists of only one kind of atom.

Compound. A compound is a pure substance which contains two or more elements chemically combined together in fixed proportions by mass.

Atom. An atom is the smallest particle of an element which can take part in a chemical reaction.

Many elements exist as more complex particles, consisting of a number of atoms. Helium exists as helium atoms. Hydrogen exists as pairs of atoms. Sulphur exists as eight-membered rings of atoms. These complex particles are called *molecules*. A molecule of an element is the smallest particle of an element which can exist independently. The smallest particle of a compound is also called a molecule. It must contain at least two atoms. We can therefore define a molecule in this way:

Molecule. A molecule is the smallest particle of an element or a compound which can exist independently.

7.2 Evidence for the existence of atoms and molecules

Dalton asked people to believe in the existence of particles which are too small to be seen, even under the most powerful microscope. It would be difficult today to find anyone who does not believe in atoms. This is because there is good scientific evidence for the existence of atoms. The Atomic Theory makes sense of many everyday observations in Chemistry.

Solution　You can drop a crystal of potassium permanganate, which is purple, into a beaker of water. You see the potassium permanganate slowly dissolve, until the whole solution is purple. It is easy to understand how this happens if you think of coloured particles splitting off from the crystal and moving about through the solution until they are evenly spread out. Figure 7.1 shows this.

Potassium
permanganate
crystal

Purple colour
spreading

Purple solution

Figure 7.1　A coloured substance dissolving

Diffusion　If a drop of the reddish–brown liquid, bromine, is put into the bottom of a gas jar, the gas jar very soon becomes filled with brown bromine vapour. You can understand this if you believe that the drop of liquid bromine contains molecules which are moving about. Given all the space in the gas jar to occupy, they move about until they are equally spread out through the gas jar as shown in Figure 7.2.

A jar of the dense, green gas, chlorine, is put underneath a jar of air, as shown in Figure 7.3, and the lids are removed. In five minutes, chlorine will have spread out evenly through both jars. The ability of gases to spread out and occupy all the space available to them is called diffusion. It is easy to see how this can happen if the gas consists of molecules of chlorine in constant motion. (Chlorine is denser than air, and you would not expect it to travel upwards.)

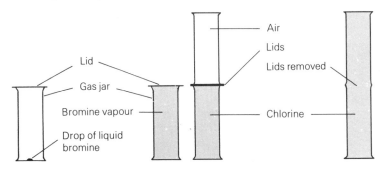

Air

Lids

Lids removed

Lid

Gas jar

Bromine vapour

Chlorine

Drop of liquid
bromine

Figure 7.2　Liquid bromine vaporising　　Figure 7.3　Diffusion of chlorine

Dilution   If you take 50 cm³ of copper sulphate solution, which is blue, and add it to 50 cm³ of water, you will find that the blue colour is half as intense as before. Add 100 cm³ of water, and you will see that the colour is a paler blue. Add the solution to 200 cm³ of water. The original 50 cm³ of copper sulphate solution is now present in 400 cm³ of solution. The colour is one eighth of its original intensity. You can explain this easily if you believe that copper sulphate consists of particles in motion. As the solution is diluted, the particles spread out through the whole of the volume, as shown in Figure 7.4

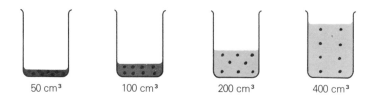

| 50 cm³ | 100 cm³ | 200 cm³ | 400 cm³ |

Figure 7.4   Dilution of a solution

Relationship between solid, liquid and gas

The Atomic Theory helps us to understand the differences between the three states of matter, solid, liquid and gas. In a solid, the atoms or molecules are arranged in a regular pattern. They cannot move out of position although they can vibrate to and fro. X-ray photographs of solids show a regular arrangement of particles. If a solid is heated, the particles gain energy. They vibrate more and more, until they break away from their regular arrangement, and move about on their own. When this happens, the solid has turned to a liquid: it has melted.

In a liquid, the molecules are free to move. Since there is no regular arrangement of molecules, the liquid has no shape of its own. A liquid assumes the shape of its container. If the liquid is heated further, the molecules will gain still more energy. Some molecules will acquire enough energy to break free from the body of the liquid: they will become a gas.

A gas occupies a very much larger volume than the liquid which evaporated to make it, e.g. 1 dm³ of water forms 1244 dm³ of steam. As the number of molecules present is the same, they must be much farther apart. There is a vast amount of space in between the molecules in a gas.

7.3 The Kinetic Theory of Gases

These ideas about gases have been developed into the *Kinetic Theory of Gases*. The word *kinetic* is Greek for *moving*. The Kinetic Theory tells us that gases consist of molecules which are in constant motion. The molecules move in straight lines, until a collision with the walls of the container or with other molecules changes their direction. A gas is mainly space. The actual volume of the molecules is tiny compared with the volume occupied by the gas. The molecules are so far apart that there is practically no force of attraction between them.

7.4 How small is an atom?

It is difficult to imagine just how small atoms are. Hydrogen atoms are the smallest. One atom of hydrogen weighs 0.000 000 000 000 000 000 000 00 17 g or 1.7×10^{-24} g. There are 600 000 000 000 000 000 000 000 or 6×10^{23} atoms in a gram of hydrogen.

Every dot of ink on this page is big enough to have a million atoms of hydrogen fitted across it, side by side. If you could count all the atoms in the dot of ink, you would find that the number was more than the population of the world. A pin head measures about 1 square millimetre. Since 10 million hydrogen atoms, side by side, measure 1 mm, the area of a pin head would hold 100 million million hydrogen atoms.

7.5 The structure of the atom

Dalton and other nineteenth century workers thought of the atom as a small sphere, that was the same all through. This idea was disproved by work done by Lord Rutherford from 1900 onwards. It was found that the atom is made up of smaller particles. There are three kinds of sub-atomic particles. The atoms of all elements are composed of just three types of particles. These are the 'building bricks' from which the whole Universe is constructed.

Table 7.1   *Sub-atomic particles*

Particle	Mass	Charge
Proton	1	+ 1
Electron	1/1850	−1
Neutron	1	0

The sub-atomic particles differ in mass and in electrical charge. *Protons* and *neutrons* have the same mass as an atom of hydrogen. We call this mass 1 atomic mass unit. An *electron* has 1/1850 of the mass of a hydrogen atom. This is so small that the electrons can be left out in adding up the mass of an atom. The mass of an atom is almost exactly equal to the mass of the protons plus the mass of the neutrons. Electrons are negatively charged, and protons are positively charged. Neutrons are neutral, uncharged particles. The charge on the electron is called 1 unit of negative charge, because it is the smallest known charge. The charge on the proton is equal and opposite, 1 unit of positive charge. Atoms are uncharged because the number of protons in an atom is exactly equal to the number of electrons.

Lord Rutherford showed that the protons and neutrons, the massive particles, are concentrated in a tiny volume in the centre of the atom. He called this the nucleus. It is the heavy part of the atom. The nucleus is minute compared with the volume of the atom. The rest of the atom is empty space, except for the presence of electrons. The electrons are a cloud of negative electricity. They circle round the nucleus at some distance from it. If the sports hall of your school were to represent the space taken up by an atom, the nucleus would be about the size of a pea. The electrons occupy the space around the nucleus by repelling the electrons of neighbouring atoms. In Physics, you may have learned that unlike charges attract. This is called electrostatic attraction. Like charges repel one another: this is called electrostatic repulsion. The negatively charged electrons of one atom repel the negative electrons of neighbouring atoms.

The electrons are negatively charged, and the nucleus is positively charged. Yet, something stops the electrons being pulled into the nucleus by electrostatic attraction. It is the fact that the electrons are moving fast in circular paths around the nucleus that prevents electrons from being sucked into the nucleus.

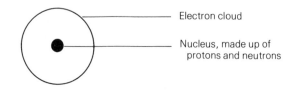

Figure 7.5   The structure of the atom (not to scale)

7.6 Symbols and formulae

Dalton used little diagrams to represent atoms. Berzelius, a Swedish chemist, had the idea of using the initial letter of an element to stand for an atom of the element, e.g. C stands for one atom of carbon. Sometimes it is necessary to use two letters: Cl stands for an atom of chlorine, Co for an atom of cobalt, Cr for an atom of chromium. He called these letters symbols. *The symbol of an element is one or more letters which stand for one atom of the element.* Sometimes, the symbols are taken from the Latin names: Pb is the symbol for lead (plumbum), and Cu is the symbol for copper (cuprum). A list of symbols is given in Table 7.2.

A symbol stands for one atom of an element. To represent two atoms, one writes a large figure 2 in front of the symbol: 2Cu means two atoms of copper. When a symbol is followed by a small number, e.g. O_2, it shows the number of atoms of the element present in one molecule. H_2 stands for a molecule of hydrogen, consisting of two atoms. S_8 stands for a molecule of sulphur. NH_3 stands for a molecule of ammonia, one nitrogen atom and three hydrogen atoms. It is the formula of ammonia.

Sometimes, a group of atoms will go through a series of chemical reactions intact. A group consisting of one oxygen atom and one hydrogen atom, OH, occurs in combination with many different metal atoms, and is called a hydroxide group. A group of one sulphur atom and four oxygen atoms, SO_4, occurs in sulphuric acid and in all metal sulphates, and is called a sulphate group. The term radical means the same as group. The formulae of some common groups are shown in Table 7.2.

The formula of a compound (or a group) is a series of symbols and numbers which tell us the ratio of atoms of each element present in the compound (or group). The formula CuO tells us that in copper oxide there is one oxygen atom for every copper atom. The formula H_2SO_4 tells us that in sulphuric acid there are two hydrogen atoms and four oxygen atoms for every sulphur atom. The formula of calcium hydroxide is $Ca(OH)_2$. The two after the bracket multiplies all the atoms in the bracket: there are two atoms of hydrogen, two atoms of oxygen and one atom of calcium. We write the formula of lead nitrate $Pb(NO_3)_2$ instead of PbN_2O_6 because it shows that the compound is a nitrate and will behave similarly to other nitrates, all of which possess the group NO_3.

Table 7.2 *Valencies of elements and groups*

The symbols and valencies of the common elements

	Element	Symbol	Valency
(a) *Metals*	Sodium	Na	1
	Potassium	K	1
	Silver	Ag	1
	Gold	Au	1 and 3
	Barium	Ba	2
	Calcium	Ca	2
	Zinc	Zn	2
	Copper	Cu	2
	Iron	Fe	2 and 3
	Lead	Pb	2
	Magnesium	Mg	2
	Aluminium	Al	3
(b) *Non-metallic elements*	Hydrogen	H	1
	Chlorine	Cl	1
	Bromine	Br	1
	Iodine	I	1
	Oxygen	O	2
	Sulphur	S	2, 4 and 6
	Nitrogen	N	3 and 5
	Phosphorus	P	3 and 5
	Carbon	C	4

The formulae and valencies of some common groups

	Group	Formula	Valency
(a) *'Metallic' group*	Ammonium	NH_4	1
(b) *Non-metallic groups*	Hydroxide	OH	1
	Nitrate	NO_3	1
	Sulphate	SO_4	2
	Carbonate	CO_3	2
	Hydrogencarbonate	HCO_3	1

The formulae of some common compounds are given in Table 7.3.

Table 7.3 *The formulae of some compounds*

H_2O	Water	CO	Carbon monoxide
NaOH	Sodium hydroxide (caustic soda)	SO_2	Sulphur dioxide
		H_2S	Hydrogen sulphide
$Ca(OH)_2$	Calcium hydroxide (Solution is lime-water)	NH_3	Ammonia
		NH_4Cl	Ammonium chloride
HCl	Hydrogen chloride (Solution is hydrochloric acid)	NH_4NO_3	Ammonium nitrate
		$(NH_4)_2SO_4$	Ammonium sulphate
HNO_3	Nitric acid	$CaCO_3$	Calcium carbonate
H_2SO_4	Sulphuric acid	$MgCO_3$	Magnesium carbonate
NaCl	Sodium chloride		
$NaNO_3$	Sodium nitrate	$Ca(HCO_3)_2$	Calcium hydrogencarbonate
Na_2SO_4	Sodium sulphate		
$CuCl_2$	Copper chloride	$Mg(HCO_3)_2$	Magnesium hydrogen carbonate
$Cu(NO_3)_2$	Copper nitrate		
$CuSO_4$	Copper sulphate	Na_2CO_3	Sodium carbonate
CuO	Copper oxide	$NaHCO_3$	Sodium hydrogencarbonate (baking soda)
$Cu(OH)_2$	Copper hydroxide		
$AlCl_3$	Aluminium chloride		
		CaO	Calcium oxide (quick-lime)
$Al(OH)_3$	Aluminium hydroxide	$ZnCl_2$	Zinc chloride
		$Zn(OH)_2$	Zinc hydroxide
Al_2O_3	Aluminium oxide	ZnO	Zinc oxide
$Al_2(SO_4)_3$	Aluminium sulphate	$ZnSO_4$	Zinc sulphate
$CuSO_4$ $5H_2O$	Copper sulphate crystals	$Na_2CO_3.$ $10H_2O$	Sodium carbonate crystals (washing soda)
CO_2	Carbon dioxide		

7.7 The ability of atoms to combine; valency

Why does hydrogen combine with chlorine to form HCl, with oxygen to form H_2O, with nitrogen to form NH_3, and with carbon to form CH_4? The reason is that different atoms can form different numbers of chemical bonds. We call the number of chemical bonds that can be formed by one atom of an element the *valency* of the element. An atom of hydrogen can form only one bond: hydrogen has a valency of one. An atom of oxygen can form two bonds: oxygen has a valency of two. Nitrogen has a valency of three, and carbon has a valency of four. Since hydrogen and chlorine both have a valency of one, they combine to form H—Cl. Since oxygen has a valency of two, one oxygen atom can combine with two hydrogen atoms giving water the formula H—O—H. Nitrogen has a valency of three: an atom of nitrogen can combine with three hydrogen atoms to give ammonia the formula

$$H-N\overset{\displaystyle H}{\underset{\displaystyle H}{}}$$

Carbon has a valency of four, and its compound with hydrogen therefore has the formula

$$\begin{array}{c} H \\ | \\ H-C-H \\ | \\ H \end{array}$$

or CH_4; it is called methane.

To understand valency better, you will have to know more about the chemical bond. What is it that holds atoms together?

7.8 What is a chemical bond?

What is it that gives atoms the ability to combine? Remember that atoms are composed of protons (which are positively charged), electrons (which are negatively charged) and neutrons (which are uncharged). The atoms of metallic elements have a desire to give away an electron or perhaps two electrons. The atoms of non-metallic elements have a desire to gain one or two electrons. An atom of sodium wants to give away one electron, and an atom of chlorine wants to gain one electron. They can both be satisfied if an atom of sodium gives one electron to an atom of chlorine.

Atoms are neutral because the number of electrons is equal to the number of protons. If a sodium atom gives away one electron, the number of protons will be one greater than the number of electrons, and the atom will become positively charged. It is then no longer a sodium atom: it has become a sodium ion. If a chlorine atom gains one electron, the number of electrons will be one greater than the number of protons, and the chlorine atom will be negatively charged. It is then no longer a chlorine atom, but a chlorine ion. It is always called a chlor*ide* ion. *An ion is an atom or group of atoms which has become charged by gaining or losing an electron (or electrons).* Since the sodium ion is positively charged and the chloride ion is negatively charged, the opposite charges attract one another, and the ions are held together by electrostatic attraction. This attraction, holding the ions together, is a chemical bond. It is not the only type of chemical bond. It is called an *ionic bond* (or an *electrovalent bond*).

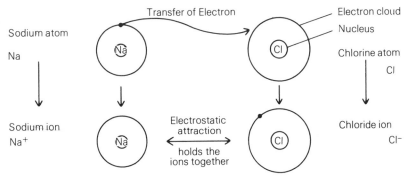

Figure 7.6   The formation of an ionic bond

If a pair of ions is formed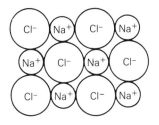

other ions will be attracted to them. Sodium ions will be attracted to the chloride ions, and chloride ions will be attracted to the sodium ions to form

This arrangement is the same in the third dimension as well. The result is a three-dimensional arrangement of alternate Na^+ and Cl^- ions. This is called an ionic lattice, and is shown in Figure 7.7 All ionic compounds are crystalline. Sodium chloride crystals are regular cubes. This is because the regular arrangement of ions results in a regular arrangement of crystal faces.

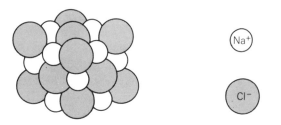

Figure 7.7   The sodium chloride lattice

Some metal ions want to give away two electrons. Calcium atoms want to give away two electrons to form calcium ions. The charge on calcium ions is $+2$ units because the number of protons is two greater than the number of electrons. A calcium atom gives one electron to each of two chlorine atoms. The chlorine atoms form chloride ions, and the compound Ca^{2+} $2Cl^-$ is formed. This is calcium chloride, which we usually write $CaCl_2$.

When magnesium combines with oxygen, a magnesium atom wants to part with two electrons, and an oxygen atom wants to find an extra two electrons. A magnesium atom therefore gives two electrons to an oxygen atom:

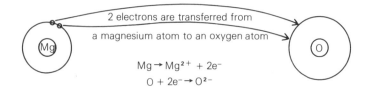

2 electrons are transferred from a magnesium atom to an oxygen atom

$$Mg \rightarrow Mg^{2+} + 2e^-$$
$$O + 2e^- \rightarrow O^{2-}$$

Figure 7.8   Two electrons being transferred from a magnesium atom to an oxygen atom

Magnesium oxide, which consists of magnesium ions, Mg^{2+}, and oxide ions, O^{2-}, is formed. The ions are held together by the electrostatic attraction between opposite charges which is called an ionic bond.

When ionic compounds are formed, the valency of the metal is equal to the number of electrons given away by one atom of the metal. The valency of the non-metallic element is equal to the number of electrons accepted by one atom of the non-metallic element. Thus, sodium has a valency of 1 because it forms Na^+ ions, chlorine has a valency of 1 because it forms Cl^- ions. Calcium, magnesium and oxygen have a valency of 2 because they form Ca^{2+}, Mg^{2+} and O^{2-} ions. *In ionic compounds, the valency of an element is equal to the charge on its ions.*

The ionic bond is not the only type of chemical bond. Two atoms of chlorine combine to form a molecule, Cl_2. Neither atom wants to give away an electron. Both atoms want to gain an electron. They do this by sharing a pair of electrons.

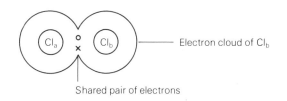

Figure 7.9   A covalent bond

The electron (o) came from Cl_a; the electron (x) came from Cl_b. Both of the chlorine atoms now share the two electrons. They have both satisfied their desire for another electron by sharing. The two atoms have to remain close together in order to share the pair of electrons. This type of chemical bond is called a *covalent bond*.

In a molecule of hydrogen chloride, an atom of hydrogen and an atom of chlorine share a pair of electrons:

(**x** = electron from H atom
o = electron from Cl atom)

Figure 7.10   Hydrogen chloride, a covalent molecule

In a covalent compound, the valency of each element is equal to the number of pairs of electrons which an atom of the element shares with other atoms. In hydrogen chloride, a hydrogen atom and a chlorine atom share a pair of electrons, and both elements have a valency of one. In ammonia, each hydrogen atom shares one pair of electrons with the nitrogen atom, and hydrogen has a valency of one. Each nitrogen atom shares three pairs of electrons with other atoms, and nitrogen has a valency of three:

(x = Electron from H atom
o = Electron from N atom)

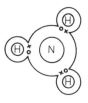

Figure 7.11   Ammonia, a covalent molecule

Ionic bonds are formed between a metallic element and a non-metallic element. A three-dimensional arrangement of ions called a *crystal lattice* is built up. All the ions are held together in a rigid, solid structure. This is why ionic compounds like sodium chloride and copper sulphate are crystalline solids with very high melting points.

Covalent bonds are formed between atoms of non-metallic elements. They are strong bonds, binding the atoms together into molecules. There are only weak forces of attraction between a molecule and its neighbours. Each molecule exists separately from other molecules, and covalent substances are often liquids or gases.

Exceptions are diamond and graphite. In these forms of carbon, large numbers of carbon atoms are bonded together by strong, covalent bonds. The atoms are arranged in a regular pattern to form giant molecules. In diamond, which is illustrated in Figure 7.12 (a), every carbon atom is bonded to four other carbon atoms, and a very strong crystal lattice is formed as a result. It makes diamond the hardest naturally occurring material. In graphite, which is shown in Figure 7.12 (b), the layers of carbon atoms can slide over one another. This is why graphite is a lubricant. The allotropes of carbon (see Chapter 6) differ in the arrangement of carbon atoms:

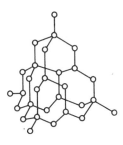

Figure 7.12   (a) Arrangement of carbon atoms in diamond

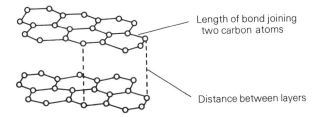

Figure 7.12   (b) Arrangement of carbon atoms in graphite

Summary   It may be useful to summarise this section on the chemical bond. In ionic bonds (or electrovalent bonds), a metallic atom gives an electron to a non-metallic atom. The metallic atom becomes a positive ion, and the non-metallic atom becomes a negative ion. There is a force of attraction between oppositely charged ions which holds them together. It is called an ionic bond. It is not the only sort of chemical bond. In the covalent bond, non-metallic atoms share a pair of electrons. They have to be close together, with their electron clouds overlapping, to share electrons. The shared pair of electrons is called a covalent bond.

Ionic substances consist of a regular arrangement of ions, and are solids. Most covalent substances consist of individual molecules, and are liquids or gases. Other covalent substances consist of giant molecules, containing thousands of atoms. These substances are solids.

129

7.9 How to work out the formulae of compounds

You have seen how ionic compounds consist of oppositely charged ions. The compound formed is neutral because the charge on the positive ion (or ions) is equal to the charge on the negative ion (or ions). In calcium chloride, $CaCl_2$, one calcium ion, Ca^{2+}, is balanced in charge by two chloride ions, $2Cl^-$. *This is how we work out the formulae of ionic compounds.*

Compound	*Copper chloride*
Ions present are	Cu^{2+} Cl^-
Now balance the charges	One Cu^{2+} ion needs two Cl^- ions.
Ions needed are	Cu^{2+} and $2Cl^-$
The formula is	$CuCl_2$
Compound	*Sodium sulphate*
Ions present are	Na^+ SO_4^{2-}
Now balance the charges	Two Na^+ are needed to balance one SO_4^{2-}.
Ions needed are	$2\,Na^+$ and SO_4^{2-}
The formula is	Na_2SO_4
Compound	*Iron(II) sulphate*
Ions present are	Fe^{2+} SO_4^{2-}
Now balance the charges	One Fe^{2+} balances one SO_4^{2-}.
Ions needed are	Fe^{2+} and SO_4^{2-}
The formula is	$FeSO_4$
Compound	*Iron(III) sulphate*
Ions present are	Fe^{3+} SO_4^{2-}
Now balance the charges	Two Fe^{3+} would balance three SO_4^{2-}.
Ions needed are	Two Fe^{3+} and three SO_4^{2-}
The formula is	$Fe_2(SO_4)_3$

You need to learn the symbols and charges of the ions in Table 7.4. Then you can work out the formula of any ionic compound.

You will notice how the compounds of iron are named iron(II) sulphate and iron(III) sulphate, to show which of its valencies iron is using in the compound. This is always done with the compounds of elements of variable valency.

130

Table 7.4 *Symbols and valencies of common ions*

Name	Symbol or formula	Valency
Hydrogen	H^+	1
Ammonium	NH_4^+	1
Potassium	K^+	1
Sodium	Na^+	1
Silver	Ag^+	1
Barium	Ba^{2+}	2
Calcium	Ca^{2+}	2
Copper(II)	Cu^{2+}	2
Iron(II)	Fe^{2+}	2
Lead	Pb^{2+}	2
Magnesium	Mg^{2+}	2
Zinc	Zn^{2+}	2
Aluminium	Al^{3+}	3
Iron(III)	Fe^{3+}	3
Hydroxide	OH^-	1
Nitrate	NO_3^-	1
Chloride	Cl^-	1
Bromide	Br^-	1
Iodide	I^-	1
Hydrogen carbonate	HCO_3^-	1
Oxide	O^{2-}	2
Sulphide	S^{2-}	2
Sulphite	SO_3^{2-}	2
Sulphate	SO_4^{2-}	2
Carbonate	CO_3^{2-}	2

To work out the formulae of covalent compounds, you need to know the symbols and valencies of the elements present. These are listed in Table 7.2. The valency is the number of electrons which an atom of the element shares in forming a compound. The method of working out the formulae is the same as for ionic compounds although here electrons are shared not given and accepted.

Compound	*Compound of carbon and hydrogen*
Symbols of elements	C　　H
Valencies (no. of shared electrons)	4　　1
Balance the electrons	One C with four e^- needs four H with one e^-.
Atoms needed	C and 4 H
The formula is	CH_4
	(This compound is called methane.)

Compound	*Compound of nitrogen and hydrogen*
Symbols of elements	N　　H
Valencies (no. of shared electrons)	3　　1
Balance the electrons	One N with three e^- needs three H with one e^-.
Atoms needed	N and 3 H
The formula is	NH_3
	(This compound is called ammonia.)

Compound	*Compound of oxygen and hydrogen*
Symbols	O　　H
Valencies (no. of shared electrons)	2　　1
Balance the electrons	One O with two e^- needs two H with one e^-.
Atoms needed	O and 2 H
The formula is	H_2O
	(This is the formula of water.)

7.10 Equations

In a chemical reaction, atoms are neither created nor destroyed. The atoms we start with are the same as the atoms we finish with, both in type and number. Old bonds are broken and new bonds are formed as the atoms enter into different arrangements, but the atoms themselves are unchanged. By using the symbols of the elements taking part in a reaction, we can show precisely what happens in a chemical reaction. We call this way of describing a reaction a chemical equation. Previously we have used word equations such as:

$$Copper + Oxygen \rightarrow Copper\ oxide$$

Now, we shall use symbols:

$$Cu + O_2 \rightarrow CuO$$

We must use O_2 to represent a molecule, not O for an atom of oxygen because the oxygen in the air, with which copper reacts, is in the form of molecules of oxygen, O_2. The two sides of an equation must be equal, and, looking at this equation to see whether the two sides are equal, we find that the left-hand side has one copper atom and two atoms of oxygen, whereas the right-hand side has one copper atom and one oxygen atom.

We must make the two sides equal. To give two oxygen atoms on the right-hand side, we need 2CuO.

$$Cu + O_2 \rightarrow 2CuO$$

We now see that the left-hand side has one copper atom, and the right-hand side has two. We put two copper atoms on the left-hand side:

$$2Cu + O_2 \rightarrow 2CuO$$

This is a chemical equation. It shows that two atoms of copper combine with a molecule of oxygen to form two molecules of copper oxide.

Magnesium and oxygen form magnesium oxide:

$$Mg + O_2 \rightarrow MgO$$

Balancing the equation (the term we use to describe making the two sides come equal) gives

$$2Mg + O_2 \rightarrow 2MgO$$

Sulphur and oxygen form sulphur dioxide

$$S + O_2 \rightarrow SO_2$$

Carbon and oxygen form carbon dioxide

$$C + O_2 \rightarrow CO_2$$

133

We can very easily put a little more information into an equation by a letter in brackets to tell us the physical state the chemical is in: (s) means solid, (l) means liquid, (g) means gas, (aq) means in aqueous (water) solution. These letters are called state symbols. Looking through the equations we have written and adding state symbols, we obtain:

$$2Cu(s) + O_2(g) \rightarrow 2CuO(s)$$
$$2Mg(s) + O_2(g) \rightarrow 2MgO(s)$$
$$C(s) + O_2(g) \rightarrow CO_2(g)$$
$$S(s) + O_2(g) \rightarrow SO_2(g)$$

This inclusion of state symbols puts in the additional information that the products copper oxide and magnesium oxide are solids and the products carbon dioxide and sulphur dioxide are gases.

Try writing equations for the reactions we carried out in Chapters 4, 5 and 6.

(1) Aluminium + Iodine → Aluminium iodide
(2) Mercury + Iodine → Mercury iodide
(3) Zinc + Sulphur → Zinc sulphide
(4) Iron + Sulphur → Iron(II) sulphide
(5) Sodium oxide + Water → Sodium hydroxide solution
(6) Magnesium oxide + Water → Magnesium hydroxide solution
(7) Calcium oxide + Water → Calcium hydroxide solution
(8) Carbon + Oxygen → Carbon dioxide
(9) Carbon dioxide + Calcium hydroxide solution → Calcium carbonate + Water
(10) Carbon + Copper oxide → Carbon dioxide + Copper
(11) Carbon + Lead oxide → Carbon dioxide + Lead
(12) Calcium carbonate → Calcium oxide + Carbon dioxide

Questions on Chapter 7

1. What do you understand by these terms: (a) element, (b) compound, (c) atom and (d) molecule?

2. Give the symbols for these elements:

Zinc	Copper	Chlorine	Nitrogen
Lead	Calcium	Oxygen	Potassium
Iron	Iodine	Sodium	Magnesium

3. What do you understand by the word atom? What three things did Dalton say cannot happen to atoms? In a chemical reaction, what happens to the atoms which take part?

4. In the square below are the names of 23 elements all written across or down – for instance, Tin. Now study the square, and see if you can find the other 22.

T	I	N	C	H	L	O	R	I	N	E	H
P	N	E	O	N	E	Z	S	N	X	X	Y
H	I	R	O	N	A	I	O	D	X	S	D
O	T	G	O	L	D	N	D	I	M	I	R
S	R	X	C	X	X	C	I	U	A	L	O
P	O	T	A	S	S	I	U	M	G	V	G
H	G	Z	R	U	X	X	M	X	N	E	E
O	E	I	B	L	C	O	P	P	E	R	N
R	N	N	O	P	I	R	O	N	S	X	X
U	X	C	N	H	X	B	A	R	I	U	M
S	X	A	L	U	M	I	N	I	U	M	X
F	L	U	O	R	I	N	E	X	M	X	X

Write down words which will fill the gaps in these passages.

5. Solids have both a definite size and a definite _____. This is because the atoms in a solid are tightly packed together. A liquid has a definite _____ but no definite _____. It fits itself into the _____ of its container. It can do this because the molecules in a liquid are free to _____. The theory which describes the behaviour of gases is called the _____ Theory of Gases. According to this theory, the molecules are always _____. This is why the molecules of a gas can spread out to occupy all the space available to them. We call this behaviour _____.

6. An atom consists of three kinds of particles. They are protons, which have a _____ charge, electrons, which have a _____ charge, and _____. The number of protons is _____ _____ the number of electrons. This is why atoms are _____. If a metal atom loses an electron, it becomes an _____ with a _____ charge. If a non-metallic atom accepts an electron, it becomes an _____ with a _____ charge. The force which holds oppositely charged particles together is called an _____ bond. When atoms share electrons, the bond is called a _____ bond.

7. Compounds which consist of ions are bonded together by _____ bonds. The arrangement of ions can be described as a _____ _____. Such compounds exist in the _____ state at room temperature. Compounds in which the atoms are bonded together by shared pairs of electrons are said to be _____ bonded. They consist of individual molecules and therefore exist in the _____ or _____ state at room temperature. Giant molecules can be built up by _____ bonding. Examples are the two allotropes of carbon, _____ and _____.

8. Give the formulae for each of these compounds:

Ammonia	Sodium hydroxide
Carbon dioxide	Sodium sulphate
Calcium oxide	Copper oxide
Zinc hydroxide	Calcium hydroxide
Copper chloride	Potassium carbonate
Sodium nitrate	Copper sulphate

9. What is the number of atoms in each of these formulae?

$NaNO_3$	$AlCl_3$	$CuSO_4.5H_2O$
Na_2SO_4	$Al(OH)_3$	$Na_2CO_3.10H_2O$
$CaSO_4$	$Pb(OH)_2$	$MgSO_4.7H_2O$
$NaNO_3$	$Al_2(SO_4)_3$	$BaCl_2.2H_2O$

Crossword on Chapter 7

Across

1 Solid form of water (3)

4 Stands for one atom, according to Dalton's Theory (6)

7 Atoms cannot be this (7)

9 These drops of liquid can be the result of sorrow or joy (5)

13 Same clue as 7 across (7)

14 A negatively charged particle (8)

16 Symbol for iron (2)

17 The smallest particle of a compound (8)

18 Symbol for nickel (2)

20 To lift up (5)

23 Symbol for magnesium (2)

24 The atoms of an element are all this (9)

26 Contains at least two elements (8)

31 Solid chemicals can cause a blockage here (5)

32 This is what atoms do with electrons in covalent bonds (5)

Down

1 Charged atoms (4)

2 The French school (5)

3 Same clue as 7 across (9)

4 Guided (7)

5 Old-fashioned word for you (2)

6 Doctor of Medicine (2)

8 A positively charged particle (6)

9 Symbol for technetium (2)

10 It evaporates quickly (5)

11 Symbol for argon (2)

12 Symbol for tin (2)

15 A way of representing a chemical reaction (8)

16 A set of symbols and numbers (7)

19 Its symbol is I (6)

21 Symbol for silver (2)

22 Symbol for silicon (2)

25 Its symbol is Ne (4)

26 Symbol for cadmium (2)

27 Alternatively (2)

28 Short for Master of Arts (2)

29 A Greek letter used in mathematics (2)

30 Domestic Science (1, 1)

Trace this grid on to a piece of paper. Then fill in the answers.

8. Water

8.1 Is water an element or a mixture or a compound?

The ancient philosophers thought the universe was made up of four elements, earth, air, fire and water. We have seen that they were wrong about air, which is a mixture of oxygen, nitrogen and small quantities of other gases. Let us now take a close look at water to see whether it can be split up if we try a number of methods. When the methods of separating mixtures in Chapter 2 were tried on water, water went through all these methods as a single substance: it cannot be a mixture. Heat does not split water up: it simply distils as one substance of boiling point 100 °C.

Another method of trying to split a substance up is to pass an electric current through the substance.

Experiment 8.1

To pass a current from a battery or a labpack through water

Figure 8.1 shows an apparatus which can be used

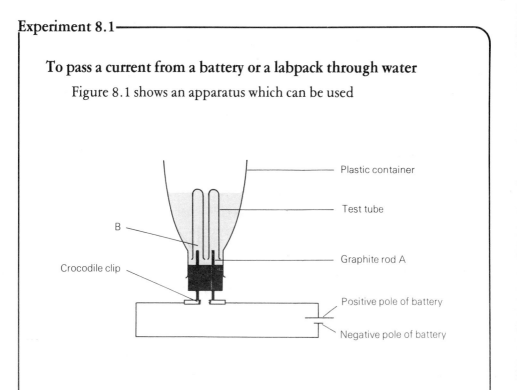

Figure 8.1   Passing an electric current through water

1. Fit graphite rods, which are called electrodes because they carry an electric current, through a rubber bung into a container. The one shown is a plastic juice bottle with 10 cm cut off the bottom. Attach crocodile clips to leads coming from a 6 V battery or a labpack adjusted to give 6–12 V.

2. Fill the vessel with distilled water. Add a few drops of dilute sulphuric acid. This helps the water to conduct electricity. Connect the crocodile clips to the graphite electrodes.

3. If you see bubbles of gas forming at an electrode, fill a test tube with water, put your thumb over the open end, and invert it over the electrode. It is better to clamp the test tube, not to rest it on the rubber bung.

4. Test any gas collected. Put your thumb under the test tube, remove it from the water, put a lighted splint into the test tube.

5. Was a gas formed at A? How did it react to a lighted splint? Was a gas formed at B? How did it react to a lighted splint?

If you have made the tests described in Experiment 8.1, you have found that the gas at A, which is connected to the positive end of the battery, is oxygen. The gas at B, which is connected to the negative end of the battery, burns with an explosive 'pop'. It is another element, and was given the name *hydrogen*, which means *water-maker* in Greek. Thus water has been split up by electricity into the elements hydrogen and oxygen. The process of splitting up by electricity is called *electrolysis*. Water is a compound of hydrogen and oxygen. You may have noticed that the volume of hydrogen produced is twice that of oxygen. Experiments on these lines have shown that the formula of water is H_2O.

8.2 Tests for water

There are many colourless liquids. You cannot assume that any colourless, odourless liquid you see is water. You need to make tests. There are many chemical reactions in which water plays a part. Here are two colourful ones, which will tell you whether a liquid contains water. They do not tell you whether the liquid is pure water or not.

Copper sulphate which contains no water is said to be anhydrous. It has the formula $CuSO_4$, and it is white. When water is added, a compound of copper sulphate and water called a hydrate is formed. It has the formula $CuSO_4.5H_2O$, and it is blue. Any liquid which contains water will turn anhydrous copper sulphate from white to blue.

Anhydrous cobalt chloride is blue. When water is added, the hydrate $CoCl_2.6H_2O$ is formed, and this is pink. When you see anhydrous cobalt chloride change from blue to pink, you know that water is present.

Having found out that the liquid contains water, you can now test to see whether it is pure water. Water has boiling point 100 °C, freezing point 0 °C, and density 1 g cm^{-3}. To find the boiling point, set up the apparatus shown in Figure 8.2. When the liquid boils, record the temperature shown by the thermometer.

To find the freezing point, you set up the apparatus shown in Figure 8.3. The mixture of ice and salt cools the water in the test tube. The temperature recorded on the thermometer falls and then remains constant when the water in the test tube starts to freeze, and stays constant all the time the liquid around it is freezing.

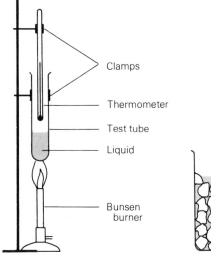

Clamps

Thermometer

Test tube

Liquid

Bunsen burner

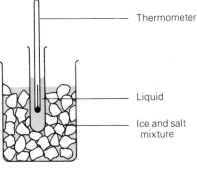

Thermometer

Liquid

Ice and salt mixture

Figure 8.2   Finding the boiling point of a liquid

Figure 8.3   Finding the freezing point of a liquid

8.3 Metals which react with water

In this section, we are going to study the reactions which take place between water and metals. One you have already met is the rusting of iron (Chapter 5). There are other metals which react with water.

Demonstration Experiment 8.2

To study the reaction between sodium and water

A safety screen should be used for this demonstration

1. Sodium is so reactive that it is kept under oil to prevent air and water from reaching it. Cut a piece of sodium the size of a pea.

2. Drop the piece of sodium into a trough of water. Observe all that happens, and test the water with litmus.

3. To test the idea that heat is generated, see what happens when sodium is not free to move. Float a piece of filter paper in the trough, and drop a piece of sodium onto it. The sodium cannot move, but water can reach it through the filter paper.

    The sodium bursts into flame, and burns with a yellow flame – or is it a gas formed during reaction that burns?

4. To test for gas, position a Pyrex tube around the sodium as shown in Figure 8.4. When the sodium has finished reacting, bring a lighted taper to the top of the tube.

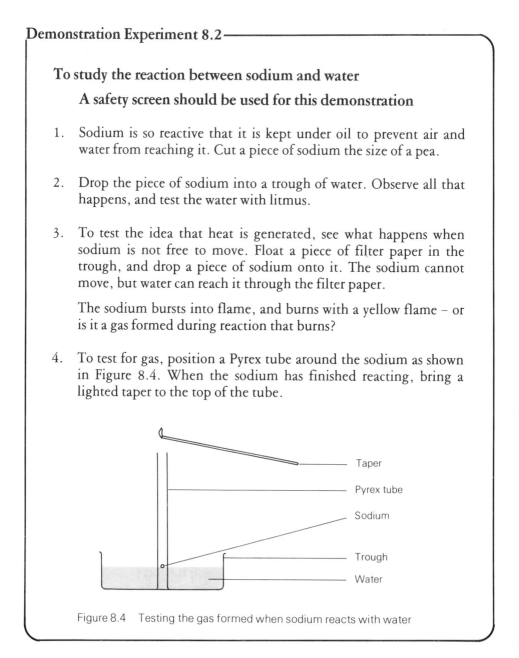

Figure 8.4   Testing the gas formed when sodium reacts with water

Sodium

Sodium is a metal. It is a much softer metal than iron, and can be cut easily with a knife. When freshly cut, sodium has a lovely silvery appearance. It soon tarnishes because sodium is a very reactive metal, and sodium oxide quickly forms on the surface.

Sodium reacts with water, and this is why sodium must be kept under a layer of oil. In Experiment 8.2, you notice that sodium floats on water: it is a very light metal. It melts into a sphere, and moves around rapidly on the surface of the water, becoming smaller and smaller until eventually it disappears. In order to melt the sodium, heat must be given out when sodium reacts with water. In order to propel sodium across the water, a gas must be formed during the reaction and act by jet propulsion. An alkaline solution is formed.

If you have tested the gas formed, as shown in Figure 8.4, you will have heard an explosive 'pop'. You have met the gas which burns with a 'pop' before: it is hydrogen, one of the elements in water. Sodium has displaced hydrogen from water. The water has become alkaline because it is now a solution of sodium hydroxide:

Sodium + Water → Hydrogen + Sodium hydroxide solution

$$2Na(s) + 2H_2O(l) \rightarrow H_2(g) + 2NaOH(aq)$$

Since sodium reacts with water to form a strong alkali, it is called an alkali metal.

Potassium

Even when it is free to move across the surface of the water, potassium bursts into flame because so much heat is given out in its reaction with water. The flame is lilac-coloured. It is really hydrogen burning with a flame coloured lilac by potassium vapour. The alkaline solution formed is potassium hydroxide:

Potassium + Water → Hydrogen + Potassium hydroxide solution

$$2K(s) + 2H_2O(l) \rightarrow H_2(g) + 2KOH(aq)$$

143

Experiment 8.4

To study the reaction of calcium with water

1. Take a 250 cm^3 beaker three quarters full of water. Add a piece of calcium. What happens?

2. Invert a test tube full of water over the piece of calcium. Collect the gas formed, as in Figure 8.5 (a). Put your thumb over the open end of the test tube while the tube is under water, as in (b). Open the tube at a flame, as in (c). What happens?

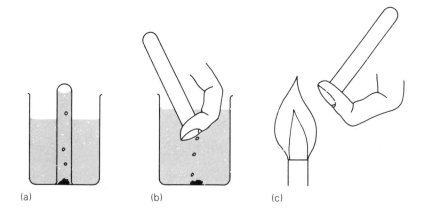

(a)   (b)   (c)

Figure 8.5   The reaction between calcium and water

3. When the calcium has reacted, what is left in the beaker? Can you name this substance?

4. Filter the contents of the beaker.

5. Take a portion of the filtrate in a test tube. Add two drops of litmus solution. What happens?

6. Take a second portion of the filtrate and blow into it through a straw. What happens? Can you name this solution?

To find out whether magnesium reacts with water

1. Sprinkle a spatula measure of magnesium powder into a 400 cm^3 beaker full of water. Cover the magnesium powder with a filter funnel, as shown in Figure 8.6. Invert a test tube of water over the neck of the filter funnel.

2. Leave to stand. Test any gas formed by opening the tube at a flame.

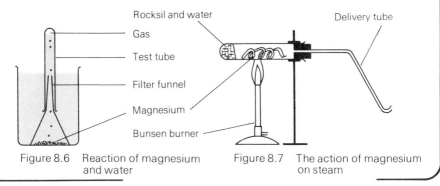

Figure 8.6   Reaction of magnesium and water

Figure 8.7   The action of magnesium on steam

To find out whether magnesium reacts with steam

1. Place loosely packed rocksil in a hard glass test tube to a depth of 3 cm. Add, by a teat pipette, as much water as the rocksil will hold (about 2 cm^3).

2. Coil a piece of magnesium ribbon, and put the coil into the middle of the tube. Fit a delivery tube and clamp as shown in Figure 8.7.

3. Wearing safety glasses, heat the magnesium. Enough heat will reach the water and turn it to steam.

4. When you see that a reaction is occurring between the magnesium and the steam, apply the Bunsen flame to the end of the delivery tube. Make a note of what you observe.

5. When the reaction is over, tip out the product formed by magnesium and steam. Add water, warm, and test with litmus.

Calcium   Calcium is a metal which is harder than sodium but softer than iron. It reacts with water, as sodium does, but less violently. It is kept in a closed bottle, but not under oil.

In Experiment 8.4, you will have seen how readily calcium reacts with water. From the tests you did, you will have recognised lime-water. It is the solution which turns milky when you breathe carbon dioxide into it. This is formed in two stages. Calcium reacts with water to produce hydrogen and a white solid. If you get hold of a piece of calcium with this white solid on the surface, it will burn your skin. The white solid is calcium oxide, or quicklime, which is so keen to get hold of water that it will take water out of your skin and so give you a burning sensation. When calcium oxide reacts with water, calcium hydroxide, slaked lime, is formed. This dissolves in water to form the solution called limewater:

$$Calcium + Water \rightarrow Calcium\ oxide + Hydrogen$$

$$Ca(s) + H_2O(l) \rightarrow CaO(s) + H_2(g)$$

$$Calcium\ oxide + Water \rightarrow Calcium\ hydroxide$$

$$CaO(s) + H_2O(l) \rightarrow Ca(OH)_2(s)$$

$$Calcium\ hydroxide + Water \rightarrow Calcium\ hydroxide\ solution$$

$$Ca(OH)_2(s) + H_2O(l) \rightarrow Ca(OH)_2(aq)$$

Calcium hydroxide + Carbon → Calcium + Water
solution   dioxide   carbonate

$$Ca(OH)_2(aq) + CO_2(g) \rightarrow CaCO_3(s) + H_2O(l)$$

Magnesium   You will find that it takes many days before Experiment 8.5 gives enough hydrogen to 'pop'. A way of speeding up the reaction is to heat it. You can do this according to Experiment 8.6. You will see a fierce glow as heat is given out when magnesium burns in steam. The gas which is formed burns at the end of the delivery tube with a blue flame. This is hydrogen. The white solid which dissolves in water to give an alkaline solution is magnesium oxide:

$$Magnesium + Steam \rightarrow Magnesium\ oxide + Hydrogen$$

$$Mg(s) + H_2O(g) \rightarrow MgO(s) + H_2(g)$$

146

To study the action of steam on iron

1. Place loosely packed rocksil in a hard glass test tube to a depth of 3 cm. Add with a teat pipette as much water as the rocksil will hold.

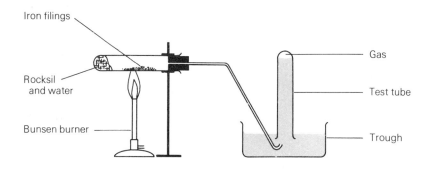

Figure 8.8   The reaction of steam and iron

2. Spread a spatula measure of iron filings in the test tube. Attach a delivery tube, and position a test tube to collect any gas formed, as shown in Figure 8.8.

3. Heat the part of the test tube containing iron. Do not heat the rocksil directly: by moving the flame to and fro, boil the water in the rocksil and keep the metal hot. In this way, a gentle flow of steam is passed over the hot metal.

4. Collect any gas formed by displacement of water.

5. With your thumb over the end of the test tube, remove it and open it at a flame. What happens?

6. As soon as you stop heating, lift the delivery tube out of the trough to avoid 'sucking back'. This is the tendency of water to travel up the delivery tube when the pressure of hot gases in the Pyrex tube drops. Cold water reaching the hot tube will crack it.

7. Observe the appearance of the iron.

Experiment 8.8

To study the reaction between zinc and steam

Carry out this experiment in exactly the same way as for iron and steam.

Iron

Iron is our most important metal. Most of the machines we use are made of iron or alloys of iron, such as steel. You know that the reaction between iron and cold water to form rust is slow. You studied this reaction in Chapter 5. In Experiment 8.7, you looked at the action of steam on iron.

You will have found that hydrogen is formed, and iron filings change from a grey colour to bluish–black. This change might remind you of heating iron in air or oxygen: it is due to the formation of iron oxide, Fe_3O_4. This is different from rust, which has the formula Fe_2O_3.

$$Iron + Steam \rightarrow Iron\ oxide + Hydrogen$$

$$3Fe(s) + 4H_2O(g) \rightarrow Fe_3O_4(s) + 4H_2(g)$$

The reaction between scrap iron and steam was until recently used as an industrial method of manufacturing hydrogen.

Zinc

If you have done Experiment 8.8, which is a study of the action of steam on zinc, you will have found that zinc is a metal of similar reactivity to iron. You will have obtained hydrogen, as before, and a coating on the zinc which is yellow when hot and white when cold. This is zinc oxide.

$$Zinc + Steam \rightarrow Zinc\ oxide + Hydrogen$$

$$Zn(s) + H_2O(g) \rightarrow ZnO(s) + H_2(g)$$

8.4 Metals which react with dilute acids

Robert Boyle worked in the seventeenth century. He was the fourteenth child of the Earl of Cork. As a member of the landed gentry, he could have measured out his life with parties and race meetings, but he preferred to spend a large part of his time in scientific research. He is famous in Physics for *Boyle's Law*, connecting the pressure and volume of a gas. In Chemistry, one of his achievements was to discover a new way of making hydrogen. Instead of reacting metals with water, he tried dilute acids. These all contain hydrogen, and they part with it more readily than water does.

148

To investigate the action of dilute hydrochloric acid on metals

1. Place a row of test tubes in a rack. Half fill them with dilute hydrochloric acid.

2. Add a piece of metal to each, labelling the tubes as you do so. *Do not* use the metals which react with cold water (sodium, potassium and calcium) as they will react too vigorously with acid. Use the metals magnesium, zinc, aluminium, tin, iron, lead and copper.

3. Observe and write down any changes which occur in the test tube. If a gas is evolved, keep your thumb over the open end of the test tube and then open it at a flame.

4. Draw up a table of your results.

Reaction of metals with hydrochloric acid

Metal	Is hydrogen evolved?	Is reaction fast or slow?	Appearance of solution

To find out whether magnesium will displace copper from a solution of copper sulphate

1. Take a test tube half filled with copper sulphate solution. Drop a piece of magnesium ribbon into it, and leave it to stand.

2. Observe any changes in the metal and in the solution.

To put metals in order of reactivity by studying displacement reactions

1. Fill a test tube half full of a concentrated solution of a compound of one metal. Hang a strip of another metal in it, as in Figure 8.9. Leave to stand for quarter of an hour.

2. Inspect carefully. Make a note of any changes in the metal and in the solution. Which metals have been displaced?

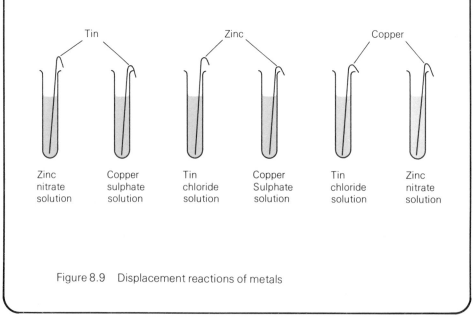

Tin	Zinc	Copper

Zinc	Copper	Tin	Copper	Tin	Zinc
nitrate	sulphate	chloride	Sulphate	chloride	nitrate
solution	solution	solution	solution	solution	solution

Figure 8.9   Displacement reactions of metals

In Experiment 8.9, you find that copper and lead have no reaction on dilute hydrochloric acid. The other metals react to give hydrogen and colourless solutions, except iron, which gives a pale green solution. The reaction of aluminium is slow, and the reaction of magnesium is fast. The reactions have similar equations.

Metal + Hydrochloric → Hydrogen + Metal chloride solution
         acid

$$Mg(s) + 2HCl(aq) \rightarrow H_2(g) + MgCl_2(aq)$$

$$Zn(s) + 2HCl(aq) \rightarrow H_2(g) + ZnCl_2(aq)$$

We have studied the reactions of metals with air, cold water, steam and dilute hydrochloric acid. The information we have obtained enables us to put the metals in order of their readiness to take part in chemical reactions. The information is summarised in Table 8.1.

Table 8.1 *Reactions of some common metals*

Element	Reaction with air	Reaction with cold water	Reaction with steam	Reaction with dilute acid
Potassium Sodium Calcium	Burn in air or oxygen	Vigorous	Dangerously fast	Dangerously fast
Magnesium		Slow	Fast	Fast
Zinc Iron		No reaction	React	Fairly fast
Tin Aluminium	Form oxide by heating in air		No reaction	Slow
Lead Copper				No reaction (except oxidising acids)

Aluminium does not show its true reactivity. The metal is surrounded by a layer of aluminium oxide which protects it from acid attack. This layer can be removed by the action of concentrated hydrochloric acid. Then the fresh aluminium underneath will show its true reactivity and place the metal just below magnesium in reactivity. Oxide films give incorrect results for other metals too. The correct order of reactivity is given in Table 8.2. Included are gold, silver, platinum and mercury, which are too expensive for us to experiment with but familiar to you outside the laboratory.

These metals are less reactive than copper. The beautiful appearance of silver and gold and the fact that they are chemically unreactive makes these metals so precious. This order, which places the metals in order of reactivity, is called the *activity series*. Hydrogen is included. Metals above hydrogen in the activity series will displace it from dilute hydrochloric and sulphuric acids. Metals below hydrogen will not displace it from dilute acids. Lead is just above hydrogen in the activity series, but in practice its reactions with dilute sulphuric and hydrochloric acids are too slow to be of use to us.

Table 8.2   *The activity series*

Potassium
Sodium
Calcium
Magnesium
Aluminium
Zinc
Iron
Tin
Lead
Hydrogen
Copper
Silver
Mercury
Platinum
Gold

When the information from chemical reactions is in doubt because the metal forms an oxide film, it is better to rely on information obtained by a physical method which measures the voltage needed to split up salts of the metals (as in Experiment 8.1 on the electrolysis of water).

Displacement reactions

When we say that one metal is more reactive than another, we mean that it is more ready to form compounds. If a metal X is more reactive, that is more ready to form compounds, than a metal Y, then X should displace Y from a solution of a compound of Y:

$$X(s) + Y \text{ compound (aq)} \rightarrow Y(s) + X \text{ compound (aq)}$$

To test this idea, take two metals widely separated in the activity series, such as magnesium and copper, and see whether magnesium will displace copper from a solution of copper sulphate. Experiment 8.10 gives details, and Experiment 8.11 extends this work.

If you have done these experiments, you will have noticed that, in Experiment 8.10, the strip of magnesium disappeared, and a reddish–brown solid appeared. You will have recognised this as copper. Magnesium has taken sulphate groups away from copper to form magnesium sulphate and copper:

$$Mg(s) + CuSO_4(aq) \rightarrow MgSO_4(aq) + Cu(s)$$

This proves that magnesium has more combining power than copper. Magnesium is higher in the activity series than copper.

You will, from your observations in Experiment 8.11, be able to place zinc, tin and copper in order of their readiness to form compounds. You will have found that:

tin does not displace zinc
tin displaces copper
zinc displaces tin
zinc displaces copper
copper does not displace tin
copper does not displace zinc.

These results place the metals in the order: zinc, tin, copper. Compare this order with their positions in the activity series.

There are very many experiments you can do with metals and compounds if you wish to extend these measurements. Two that give impressive results are the action of a strip of copper on silver nitrate solution and a strip of zinc on lead nitrate solution. The order we obtain as a result of all these displacement reactions is the same as that in Table 8.2, the activity series of the metals, which was drawn up from the reactions with water, steam and dilute acids.

8.5 Hydrogen

The action of metals on dilute acids gives us a convenient way of preparing hydrogen. Experiment 8.12 gives details which will enable you to make hydrogen and test it. If you think about it, you may be able to suggest what will be formed when hydrogen burns. If your teacher demonstrates Experiment 8.13, you will find out whether you are right. Being a gas, hydrogen diffuses. The molecules of a gas are in constant motion, and spread out to occupy all the space available to them. In Experiment 8.14, there is a competition between hydrogen and air to see which diffuses faster.

Experiment 8.12

Laboratory preparation of hydrogen

1. Set up the apparatus shown in Figure 8.10.

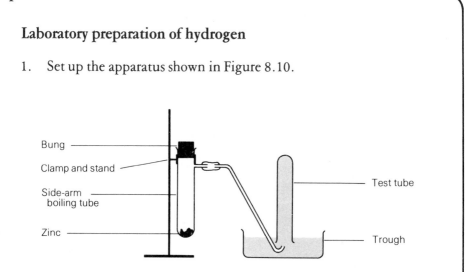

Figure 8.10   Apparatus for preparing hydrogen

2. Light some Bunsen burners on the side benches. Keep the Bunsens away from your apparatus.

3. Add 15 cm³ dilute sulphuric acid to the zinc. Stopper. Note what happens.

4. Add 1 cm³ copper sulphate solution. Stopper. Note what happens.

5. Collect the gas after it has been bubbling for half a minute.

6. Remove a test tube full of gas, closing the end with your thumb. Open it near a Bunsen flame. What happens?

7. Note the colour and smell of the gas. Collect two more test tubes of gas, and insert corks while the open ends are under water.

154

8. Is hydrogen denser or less dense than air? Put a test tube of dry air above a corked test tube of hydrogen, as shown in Figure 8.11 (a). Remove the cork, and hold the mouths of the two test tubes together while you wait ten seconds. Then take the tubes apart and immediately apply the ends to a flame. Where is the hydrogen now?

9. Repeat (8), this time with the test tube of hydrogen on top of the test tube of dry air, as in Figure 8.11 (b).

What do you observe in the dry test tube after a mixture of hydrogen and air has burned in it?

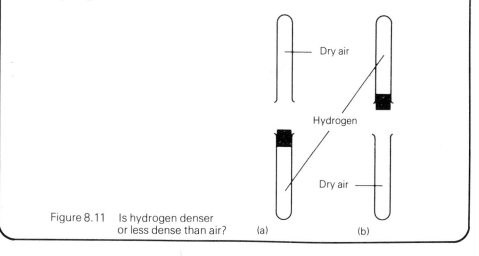

Figure 8.11 Is hydrogen denser or less dense than air?   (a)   (b)

Figure 8.12   Scientists releasing meteorological balloon

What is formed when hydrogen burns in air?

1.   Assemble the apparatus for generating hydrogen shown in Figure 8.13. Attach a U tube containing the drying agent anhydrous calcium chloride to the generator so that the hydrogen will be dry. Attach a delivery tube to the U tube. Have the side-arm tube fitted with a thistle funnel standing by.

2.   Lead the gas under water and collect in test tubes as shown in Figure 8.10. Test the gas at a flame. When there is no 'pop', but the hydrogen burns quietly, all the air has been displaced from the apparatus, and it is safe to light hydrogen at a jet.

3.   Connect the jet as shown in Figure 8.13, and light the hydrogen coming from the jet.

4.   Collect the products formed when hydrogen burns by drawing air through a thistle funnel over the hydrogen flame into a side-arm tube, surrounded by ice and water and connected to a suction pump. Let hydrogen burn for fifteen minutes, being careful to keep up a good stream of hydrogen by adding more acid to the generator if required.

5.   At the end of fifteen minutes, you will notice that a colourless liquid has condensed in the side-arm tube. Since it looks very much like water, it is logical to test for water. Two quick tests to

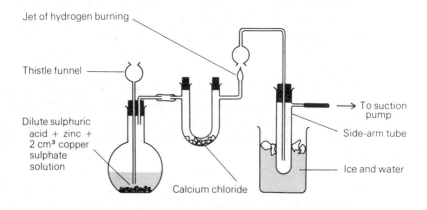

Figure 8.13   Combustion of hydrogen

find out whether the liquid contains water are: (1) Does it turn white anhydrous copper sulphate blue? (2) Does it turn blue anhydrous cobalt chloride pink? If you find that the liquid contains water, the next step is to find out whether it is pure water. Pure water freezes at 0 °C, and boils at 100 °C. You find the melting point and boiling point by the methods illustrated in Figures 8.2 and 8.3.

If you have done Experiment 8.12, you will have found that hydrogen is a colourless, odourless gas which is insoluble in water. It burns in air with an explosive 'pop'. You will have found that whereas in part 8 of this experiment hydrogen rises, in part 9 it stays at the top, showing that hydrogen is less dense than air. If you leave the test tubes in contact for a longer time, say two minutes, there will be time for diffusion to occur. Hydrogen and air will become spread evenly through the two test tubes. You should have observed condensation in the dry test tube after a mixture of hydrogen and air has burned in it. Experiment 8.13 takes this point further.

On burning hydrogen, as in Experiment 8.13, and testing as described, you will have found that the liquid formed is water. Hydrogen burns in air to form water:

$$\text{Hydrogen} + \text{Oxygen} \rightarrow \text{Water}$$

$$2H_2(g) + O_2(g) \rightarrow 2H_2O(l)$$

For a long time, water was thought to be an element. This experiment, which shows how it is formed by the combination of two elements, disproved that idea. As we have already found out, the name *hydrogen* means *water maker*, and this name was given to the element as a result of the combustion which you have just studied. You may remember that in Experiment 8.1, water was electrolysed to give hydrogen and oxygen. Water is a spectacular example of a compound with properties differing from its elements. Hydrogen, the gas which burns with a 'pop', and oxygen, the gas which supports life, combine to form the liquid water.

In Experiment 8.14, you will see the coloured liquid shoot out of the open end of the manometer. A manometer measures pressure, and this movement means that something is pressing down on the porous pot end of the tube. You have just introduced hydrogen into the beaker surrounding the porous pot: it must have passed into the porous pot to increase the pressure of gas inside the porous pot. Hydrogen diffuses into the porous pot faster than air diffuses out. Hydrogen is the lightest of gases, and diffuses faster than all others.

Experiment 8.14

To show the diffusion of hydrogen

1. Set up the flask and thistle funnel apparatus for generating hydrogen shown in Figure 8.13. Attach a delivery tube to direct hydrogen upwards into a beaker as shown in Figure 8.14. The beaker surrounds a porous pot fitted with a manometer, a U shaped piece of glass tubing containing a coloured liquid.

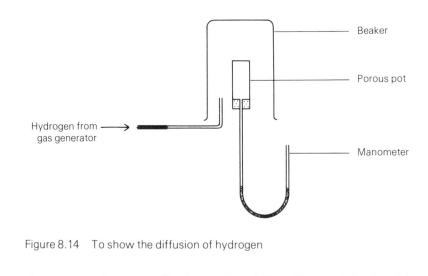

Figure 8.14 To show the diffusion of hydrogen

2. Get up a good stream of hydrogen by adding dilute sulphuric acid and a little copper sulphate solution to zinc. Then introduce the delivery tube into the beaker surrounding the porous pot.

Carbon and hydrogen are non-metallic elements with the ability to combine with oxygen. We saw in Chapter 6 that carbon will reduce some metal oxides to the metal and carbon dioxide. In the next experiment, you can find out whether hydrogen will reduce metal oxides to the metal.

Will hydrogen reduce metal oxides?

1. Put loosely packed rocksil to a depth of 3 cm in a Pyrex test tube. With a teat pipette, add as much water as it will hold.

2. Place a piece of steel wool in the middle of the test tube, and position it as shown in Figure 8.15.

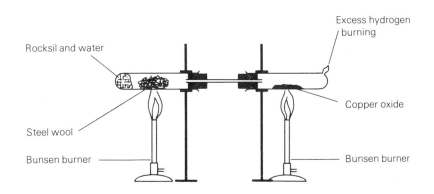

Figure 8.15   Testing hydrogen as a reducing agent

3. Take a hard glass test tube with a hole in the end, spread a spatula measure of copper oxide in the middle of the tube, and position it as in the diagram.

4. Heat the steel wool. The water will boil, and a stream of steam will pass over the heated iron. Light the hydrogen, which is produced by the action of steam on heated iron, as it comes out of the hole in the second test tube. At the same time, heat the copper oxide.

5. Observe carefully whether there is a reaction between hydrogen and heated copper oxide. If the reaction

$$\text{Hydrogen} + \text{Copper oxide} \rightarrow \text{Copper} + \text{Water}$$

$$H_2(g) + CuO(s) \rightarrow Cu(s) + H_2O(g)$$

takes place, you will see the unmistakable colour of copper and you may see some water condensing in the second test tube.

In fact, hydrogen does react with copper oxide to form copper and water. Addition of oxygen to a substance is called oxidation; addition of hydrogen or subtraction of oxygen is called reduction. In the reaction,

$$\text{Copper oxide + Hydrogen} \rightarrow \text{Copper + Water}$$

$$CuO(s) + H_2(g) \rightarrow Cu(s) + H_2O(g)$$

with Reduction indicated over the left side and Oxidation indicated under the reaction.

copper oxide loses oxygen: copper oxide is reduced. Hydrogen gains oxygen: hydrogen is oxidised. Copper oxide is an oxidising agent; hydrogen is a reducing agent.

If you repeat the experiment with other metal oxides, you find that reduction does not always occur. It is only metals low in the activity series that have oxides which can be reduced by hydrogen. These are the metals which will not displace hydrogen from acids.

Industrial manufacture of hydrogen

Petroleum oil. The petroleum industry is a big source of hydrogen. Crude petroleum oil consists of a mixture of many compounds of carbon and hydrogen called hydrocarbons. The compounds are separated by distillation in petroleum refineries. The heavy fuel oils and lubricating oils are *cracked* – that is, split up by heating – to form more of the lighter fractions, such as petrol for cars. In this process, some hydrogen is split off. It is collected, stored and sold.

Electrolysis. You saw how hydrogen is formed when water is electrolysed (split up by an electric current). Many solutions give hydrogen on electrolysis. Brine (sodium chloride solution) is electrolysed in large quantities to obtain chlorine for bleaches, etc., and hydrogen is the other product. It is collected, stored and sold.

Industrial uses of hydrogen

Manufacture of ammonia. Hydrogen is used in the manufacture of ammonia. Nitrogen and hydrogen combine under pressure in the presence of the catalyst, iron, to form ammonia. This is an important reaction as ammonia is used in the manufacture of fertilisers. Since ammonium salts are soluble, they can carry the nitrogen they contain into the soil and into the roots of plants.

Thus they replenish the nitrogen content of the soil and nourish the crops. Ammonia is important for another reason. It can be oxidised to nitric acid, a chemical of enormous importance, for example in the preparation of explosives like nitroglycerine and trinitrotoluene (TNT).

Reducing agent. Hydrogen is used as a reducing agent in the extraction of metals from their ores, for example tungsten from tungsten oxide and molybdenum from molybdenum oxide.

Fuel. Hydrogen is used as a fuel. It burns to form the harmless product water. Two gases which contain hydrogen are coal gas and water gas. Coal gas is made by destructively distilling coal in the absence of air (Chapter 6). Water gas is a mixture of hydrogen and carbon monoxide made by passing steam over red hot coke. The use of these gases as fuels has decreased since North Sea gas became available.

Hydrogenation of oils. Oils such as olive oil, whale oil and peanut oil can be converted into hard fats such as margarine by reaction with hydrogen in the presence of heated nickel. The process is called *hydrogenation*. Fats formed by hydrogenation of oils can be sold for a higher price than oils because they form a substitute for butter, which is not produced in sufficient quantity to satisfy our needs.

Balloons. Hydrogen balloons are sent up into the atmosphere by meteorologists (people who study the weather). Figure 8.12 shows scientists from the Meteorological Office Research Station releasing a hydrogen balloon. The balloon ascends carrying equipment which will record information which the scientists will later study.

8.6 Substances dissolved in water

Water is a neutral substance. It is a good solvent for a large number of materials. To obtain pure water, free from dissolved impurities, distillation is employed (see Figure 2.8). In the following experiments, you can find out whether tap water contains dissolved solids and gases. The gases which water is most likely to dissolve are the gases which make up the air. These are oxygen, nitrogen, and carbon dioxide. You can do an experiment to find out whether air is driven out of water by boiling, and, if so, how much.

To find out whether tap water contains dissolved solids

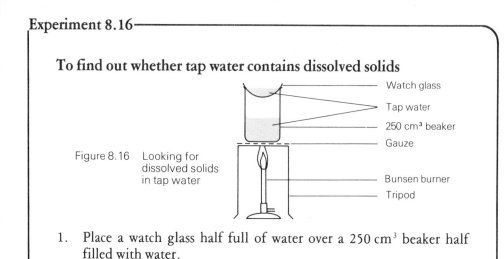

Figure 8.16   Looking for dissolved solids in tap water

Watch glass
Tap water
250 cm³ beaker
Gauze
Bunsen burner
Tripod

1. Place a watch glass half full of water over a 250 cm³ beaker half filled with water.

2. Heat the beaker of water, as shown in Figure 8.16. Steam from the water in the beaker will gently heat the water in the watch glass and cause this water to evaporate.

3. When all the water has evaporated, look at the watch glass and make a drawing of the solid in it. Why do you think it was deposited in this pattern?

To find out whether tap water contains dissolved gases

1. Take a boiling tube three quarters full of tap water. Put it into a 250 cm³ beaker half full of water, as in Figure 8.17. Heat the beaker of water.

2. Make a diagram of what you see on the sides of the boiling tube after heating.

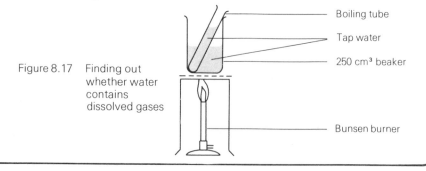

Figure 8.17   Finding out whether water contains dissolved gases

Boiling tube
Tap water
250 cm³ beaker
Bunsen burner

What volume of air is driven out of tap water on boiling?

1. Fill the 2 dm³ round-bottomed flask and the delivery tube shown in Figure 8.18 completely with water. Turn off the water tap, tighten the screw clip, and disconnect the rubber tubing from the tap.

2. Heat the water in the flask and collect the air driven out in a graduated tube full of water. When no further increase in the volume of air can be seen, stop heating.

3. Measure the volume of air collected.

4. Allow the water in the round-bottomed flask to cool. Pour some of it into a 100 cm³ graduated tube until the tube is three quarters full. Read the volume of water.

5. Invert the graduated tube in a deep vessel of water as described in Experiment 5.13, and read the volume of air at atmospheric pressure.

6. Stopper the graduated tube and shake for five minutes.

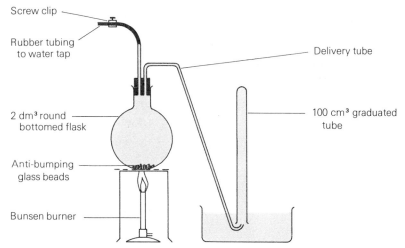

Figure 8.18   Finding the volume of air dissolved in water

7. Open the tube under water. You will see water rising into the tube to take the place of the air which has dissolved. Measure the volume of air at atmospheric pressure again.

8. Subtract to find the volume of air which has dissolved in 75 cm³ of water. What volume of air would dissolve in 2 dm³ water? Compare this figure with the volume of air driven out of 2 dm³ water on heating.

If you have done Experiments 8.16, 8.17, and 8.18, you will have found that water contains dissolved air. At 15 °C, the volume of air in 2 dm^3 of water is 43 cm^3. Your result should be close to this figure. The fact that air dissolves in water is of vital importance to fish and other aquatic animals, who need this dissolved air to live. We too depend on the solubility of air when the air we inhale into our lungs dissolves in our blood stream and is therefore able to oxidise food materials in the blood, with the liberation of energy.

Types of natural water

Rainwater. Rainwater is the purest naturally occuring form of water. It dissolves the gases of the air through which it falls, oxygen, nitrogen and the weakly acidic gas, carbon dioxide. Rainwater is therefore a weakly acidic solution. In industrial cities, rainwater dissolves sulphur dioxide, which forms the much stronger acids sulphurous acid and sulphuric acid. The city rainwater is therefore more acidic and is capable of slowly dissolving building materials such as limestone and concrete.

Spring water. Rainwater percolates through the soil until it meets a layer of rock which it cannot penetrate. It then trickles along the rock layer until it finds a crack in the earth above it, through which it can surface as a spring. Percolation removes suspended matter such as bacteria so that the water is usually drinkable. During its course along the underground rocks, various minerals dissolve in the spring water. The water of certain springs, e.g. Bath and Harrogate, contains salts such as magnesium sulphate and sodium sulphide which are helpful to people with some illnesses, and people go to these spas to drink or bathe in the water.

River water and lake water. River water and lake water contain different kinds and quantities of dissolved matter, depending on the types of rock over which the water has flowed. The salts usually found in river and lake water are mainly the sulphates, chlorides and hydrogencarbonates of calcium and magnesium. The main ones are calcium sulphate, dissolved when water trickles over rocks containing gypsum, $CaSO_4.2H_2O$, which is slightly soluble in water, and calcium hydrogencarbonate. When water permeates chalk or limestone, although these minerals are insoluble in water, they dissolve in water which contains dissolved carbon dioxide to form a solution of calcium hydrogencarbonate.

Calcium + Water + Carbon → Calcium
carbonate   dioxide·   hydrogencarbonate
                              solution

$$CaCO_3(s) + H_2O(l) + CO_2(g) \rightarrow Ca(HCO_3)_2(aq)$$

Seawater. Seawater contains up to 3.6% of dissolved solids such as sodium chloride, magnesium sulphate, magnesium chloride, potassium bromide.

Drinking water. Water is purified to make it fit for drinking by filtering it to remove solid matter and bacteria. A filter bed is shown in Figure 8.19. The water coming from it is tested and then chlorinated.

You can make a model water filter bed from a plastic juice bottle, as shown in Figure 8.20.

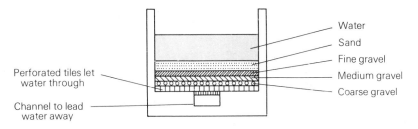

Figure 8.19   A water filter bed

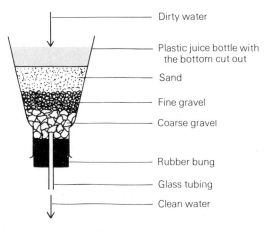

Figure 8.20   A model water filter bed

The water cycle. We have looked at the courses taken by rainwater falling to the ground and percolating through the ground to become spring water or to become river water or lake water. There are three ways in which the supply of water in the atmosphere is replenished. Water vapour passes up into the atmosphere from below by evaporation from lakes, rivers and seas. The process of transpiration in plants draws water from the soil through the roots and allows water to evaporate through the leaves. In respiration, animals burn up starchy food materials in their bodies and breathe out air containing much water vapour. Thus, respiration, transpiration and evaporation send water vapour into the upper atmosphere. When the air becomes saturated with water vapour, rain clouds form. When the clouds cool, water vapour condenses and falls to earth as rain.

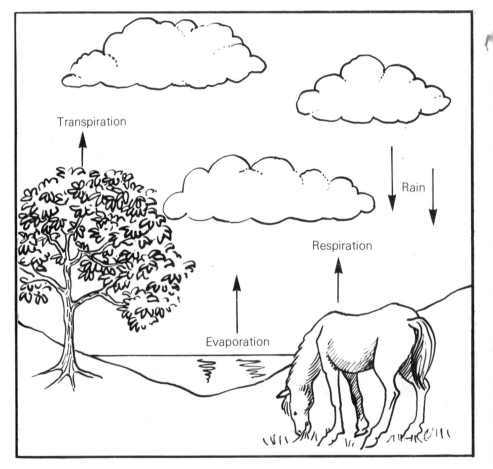

Figure 8.21 The water cycle

8.7 Soap

One reason why we need water is for washing. We keep a number of chemical manufacturers in business making soaps and detergents for our use. Soap is made by a process called *saponification*, in which fats and strong alkali are boiled together. Experiment 8.19 is a method for making soap in the laboratory.

Fats are made of glycerol combined with an acid. In soap making, these two parts are split up by sodium hydroxide to give glycerol and the sodium salt of the acid. One fat is glyceryl stearate; on saponification, it gives glycerol and sodium stearate.

Glycerol stearate + Sodium hydroxide → Glycerol + Sodium stearate

$$
\begin{array}{l}
CH_2OCOC_{17}H_{35} \\
| \\
CH\ OCOC_{17}H_{35}(l)\ +\ 3NaOH(aq) \rightarrow \\
| \\
CH_2OCOC_{17}H_{35}
\end{array}
\qquad
\begin{array}{l}
CH_2OH \\
| \\
CH\ OH(l)\ +\ 3C_{17}H_{35}COONa(aq) \\
| \\
CH_2OH
\end{array}
$$

Sodium stearate is soap. There are other soaps, sodium salts of other acids.

When you wash your hands with soap, the $C_{17}H_{35}$ hydrocarbon part is attracted to the grease on your hands, and the COONa sodium salt part is attracted to the water. The soap molecule forms a link between the grease on your hands and the water, even though grease and water do not normally mix. The grease floats away in the water, and takes solid particles of dirt with it.

Soap lathers easily in distilled water, but sometimes it is hard to get a lather in tap water and we would call this sort of tap water *hard*. Distilled water is called *soft* water. It contains very little dissolved matter, whereas tap water often contains dissolved solids that make the water hard. It is the calcium and magnesium salts in tap water which make it hard, and turn the soap into a scum instead of a lather. In some parts of the country however tap water is quite soft.

To make a bar of soap

1. Put 25 g of a mixture of 25% vegetable oil and 75% dripping with 0.25 g of bar soap into an evaporating basin.

2. Set the evaporating basin over a beaker of water as shown in Figure 8.22 (a), and boil the water.

3. Gradually add 10 cm³ of 10% sodium hydroxide solution, stirring all the time.

4. Continue to heat and stir for about thirty minutes, until the whole mass in the dish has become emulsified and the mixture stiffens and sticks to the rod.

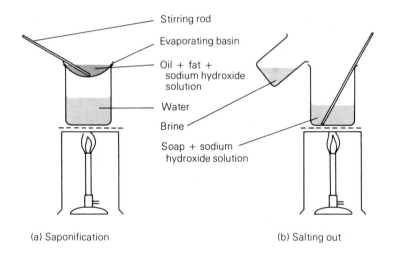

Stirring rod

Evaporating basin

Oil + fat + sodium hydroxide solution

Water

Brine

Soap + sodium hydroxide solution

(a) Saponification   (b) Salting out

Figure 8.22   Making soap

5. Cool. Scrape into a 250 cm³ beaker, and add 30 cm³ of hot water.

6. Heat over a low flame, with stirring, for 30 minutes. When the contents of the beaker appear to be a thick, even paste, add saturated brine, as in Figure 8.22 (b). Keep stirring until the soap will break quickly and evenly from the surface of a spatula dipped into it.

7. Allow to stand overnight. The soap separates as a solid layer on top of the brine. The action of the brine in making the soap solidify is called salting out. Remove the soap and dry it with a paper towel. Add perfume and colouring if you wish, and shape the soap into a bar. Leave it to dry.

8. Do three tests with the soap you have made.

   (a) Wash your hands with it – not your face, as the soap may still contain some alkali.

   (b) Shake a piece of soap in a test tube with distilled water.

   (c) Shake a piece of soap in a test tube with tap water.

Soap is sodium stearate. It reacts with calcium and magnesium compounds to form an insoluble scum of calcium stearate or magnesium stearate. When all the calcium and magnesium have been used up in the formation of scum, the soap can then produce a lather. This is an expensive way of getting a lather, and other methods of removing calcium and magnesium compounds would be an advantage.

The chief compounds present in tap water are calcium and magnesium sulphates and chlorides and calcium and magnesium hydrogencarbonates. In order to soften water, that is make it easy to get a lather from soap, these salts must be removed.

8.8 Methods of softening hard water

Temporary hardness

Boiling. Temporary hardness is hardness which can be removed by boiling. This type of hardness is caused by the presence of calcium and magnesium hydrogencarbonates. These compounds are formed when the carbonates dissolve in water containing dissolved carbon dioxide:

$$CaCO_3(s) + H_2O(l) + CO_2(g) \rightarrow Ca(HCO_3)_2(aq)$$

This process is reversed by boiling:

Calcium → Calcium + Carbon + Water
hydrogencarbonate carbonate dioxide
solution

$$Ca(HCO_3)_2(aq) \rightarrow CaCO_3(s) + CO_2(g) + H_2O(l)$$

Calcium hydrogencarbonate decomposes to give insoluble solid calcium carbonate, carbon dioxide gas and water, which is now soft water. The deposit of calcium carbonate can be seen on the inside of a kettle and is called *scale* or *fur*. It can be a problem when it builds up a thick layer inside hot water pipes.

Permanent hardness

Permanent hardness is hardness which cannot be removed by boiling. It is caused by the presence of calcium and magnesium chlorides and sulphates, which are not decomposed by heat. Various methods can be used to soften permanently hard water.

Distillation. Distillation will remove all dissolved matter. This is too expensive a method for domestic and industrial use.

Soap. Addition of soap will eventually produce a lather, but this is a wasteful method.

Washing soda. Addition of washing soda, sodium carbonate crystals, $Na_2CO_3.10H_2O$, will soften water. Calcium and magnesium present as soluble compounds are precipitated as insoluble carbonates.

Calcium sulphate + Sodium carbonate → Calcium carbonate + Sodium sulphate

$$CaSO_4(aq) + Na_2CO_3(aq) \rightarrow CaCO_3(s) + Na_2SO_4(aq)$$

Exchange resins. There are materials called ion exchange resins which will take ions of one element out of its compounds and replace them with ions of another element. Manufactured ion exchange resins are called permutits. A permutit water softener is a compound of sodium and permutit. The way it works is shown in Figure 8.23. Hard water passes through the column, and an exchange takes place:

Calcium ions + Sodium permutit → Sodium ions + Calcium permutit

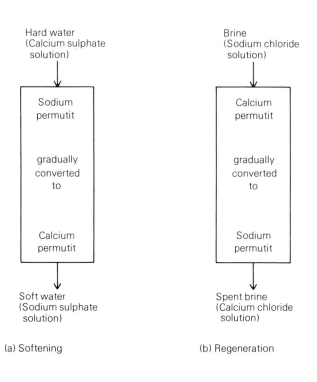

Figure 8.23   A permutit water softener

When all the sodium ions combined in the permutit have been replaced by calcium and magnesium ions, the permutit cannot go on softening water. It must be regenerated. Brine, a strong solution of sodium chloride, is passed through the column, and another exchange takes place:

Calcium permutit + Sodium ions → Sodium permutit + Calcium ions

A diagram of a domestic water softener is shown in Figure 8.24. A model of a domestic water softener is shown in Figure 8.25.

The four methods of softening permanently hard water will also soften temporarily hard water. You can do Experiment 8.18 to compare different methods of softening water.

171

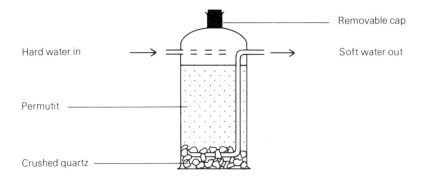

Figure 8.24   A domestic water softener

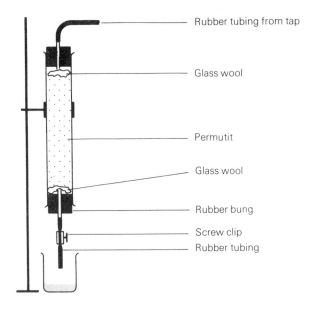

Figure 8.25   A model water softener

Experiment 8.20

To compare methods of softening tap water

1. *Tapwater.* Measure 20 cm^3 of tap water in a measuring cylinder, and pour it into a boiling tube. Add soap solution drop by drop from a teat pipette, with shaking, until a lather is formed which will last for ten seconds. Record the number of drops.

172

2. *Boiling.* Boil some tap water in a beaker for five minutes. Cool. Take 20 cm^3 in a measuring cylinder, tip into a boiling tube and again find the number of drops of soap solution needed to give a lather. Write down the number.

3. *Washing soda.* Measure 20 cm^3 of tap water from a measuring cylinder into a boiling tube. Add a crystal of washing soda, sodium carbonate, and shake to dissolve it. Again, find the number of drops of soap solution needed to give a lather.

4. *Distillation.* Distil some tap water as shown in Figure 2.8. Test 20 cm^3 of the distillate with soap solution as before.

5. *Permutit.* Pass some tap water through a model of a domestic water softener. This is easy to make if you have a wide (3 or 4 cm diameter) glass tube which is 25–30 cm long. Figure 8.25 shows how to pack the tube with permutit resting on glass wool. Run tap water in slowly so that a slow trickle of water comes out of the bottom of the water softener. Collect it in a beaker, and measure 20 cm^3 of it from a measuring cylinder into a boiling tube. Test with soap solution as before.

6. Now compare your results. Which samples of water needed the least soap solution to form a lather? Is there a high content of calcium and magnesium hydrogencarbonates (which are removed by boiling) in tapwater? Is the water still hard when these compounds have been decomposed? Which method of softening worked best for you? You can find the ratio of temporary to permanent hardness from your results:

If number of drops of soap solution needed by tap water $= a$,
and number of drops of soap solution needed after boiling $= b$,

$$\text{permanent hardness} = b$$

$$\text{temporary hardness} = a - b$$

Then $\dfrac{\text{temporary hardness}}{\text{permanent hardness}} = \dfrac{a - b}{b}$

There will be different ratios in different parts of the country.

Detergents Soapless detergents are made in a similar way to soaps, from fats and oils, but concentrated sulphuric acid is used instead of sodium hydroxide solution. They clean in a manner similar to soaps, by forming a bridge between grease and water, enabling the grease to emulsify (mix) with water. The difference is that detergents do not form a scum even in hard water because the calcium compounds of soapless detergents are soluble.

Advantages of hard water

For drinking purposes, hard water is better than soft water. The calcium content helps to form strong teeth and bones. When water is passed through lead pipes, if the water is very soft, some lead will dissolve, and may give rise to lead poisoning. If lead dissolves in hard water, it immediately forms insoluble lead sulphate or lead carbonate, which coats the inside of the lead pipe and so stops any further reaction between the water and the metal.

For a number of industrial uses, hard water is advantageous. The brewing industry and the tanning industry both use hard water.

8.9 Stalactites and stalagmites

You have read how, in limestone regions, water contains dissolved calcium hydrogencarbonate. In an underground cavern, there will be a slow trickle of water running from the roof to the floor. Imagine a drop of water becoming isolated from the main stream. With air all round it, the water evaporates, and the salt dissolved in the drop of water is left behind as a tiny speck of calcium carbonate. Another drop of water evaporates, and another speck of calcium carbonate is formed. Over a period of hundreds of years, the specks of calcium carbonate accumulate and form a cone hanging down from the roof. This is called a *stalactite*. If evaporation takes place from the floor of the cave, a pillar can build up slowly from the floor. This is called a *stalagmite*. In some places, stalactites and stalagmites have met to form complete pillars (see Figure 8.26).

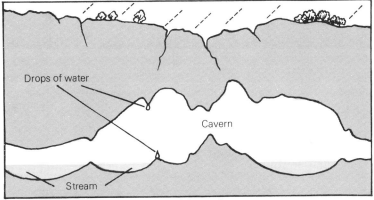

(a) Drops of water evaporate to leave calcium carbonate

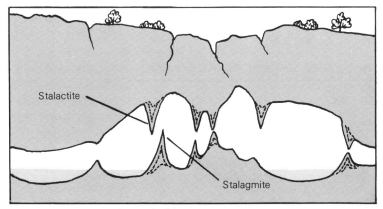

(b) Specks of calcium carbonate build up

Figure 8.26   Formation of stalactites and stalagmites

Questions on Chapter 8

1. Describe in detail what you see when:

    (a) a piece of sodium is dropped into a trough of water,

    (b) steam is passed over heated magnesium,

    (b) a strip of zinc is placed in a solution of copper sulphate.

    Explain what chemical reactions occur.

2. (a) Name three metals which react with cold water.

    (b) Name three metals which react with steam.

    (c) In each case, say what products are formed in these six reactions.

    (d) Name three metals which are resistant to attack by air, water and chemicals.

3. (a) Why do space ships carry liquid hydrogen and liquid oxygen?

    (b) Why is hydrogen used in meteorological balloons?

    (c) What is the action of hydrogen on heated copper oxide?

    (d) What is the reason for converting hydrogen to ammonia?

4. Three metals, X, Y and Z, form oxides XO, YO and ZO. Hot, powdered Y will remove oxygen from XO but not from ZO. Which is the correct order of reactivity of the elements, starting with the most reactive?

    (a) X, Z, Y (b) X, Y, Z (c) Z, Y, X (d) Y, X, Z
    (e) Z, X, Y.

5. Briefly describe how you would separate copper from a mixture of copper powder and magnesium powder.

6. (a) Iron and zinc both react with steam when heated: they are metals of similar reactivity. Describe an experiment you can do to find out which is the more reactive, iron or zinc.

    (b) Copper and silver do not react with steam or dilute hydrochloric acid: they are metals of similar reactivity. Explain how you could find out which of these two is the more reactive.

7. Write down the missing words in this passage.

Temporarily hard water contains _____, which when heated becomes _____. This process occurs naturally in limestone caves where _____ is deposited as _____ on the roof and _____ on the floor. The gas given off is _____. Temporarily hard water can be softened by _____. Permanently hard water contains magnesium _____, and can be softened by adding _____.

8. (a) Explain the difference between the action of soap on hard water and soft water.

(b) The salts which make water hard can be removed by distillation. Draw a diagram of the apparatus you would use to distil water.

(c) Describe how you would test the water before and after distillation to see whether it has been softened.

9. (a) For what purpose is soft water better than hard water?

(b) When is it better to use hard water rather than soft water?

(c) What two substances come out of solution when temporarily hard water is boiled?

(d) What would happen if a piece of stalactite were dropped into dilute hydrochloric acid?

10. (a) Name two compounds which cause temporary hardness in water. Briefly describe two methods for softening this water.

(b) Name two compounds which cause permanent hardness in water. Briefly describe two methods by which this water may be softened.

Trace this grid on to a piece of paper, and then fill in the answers.

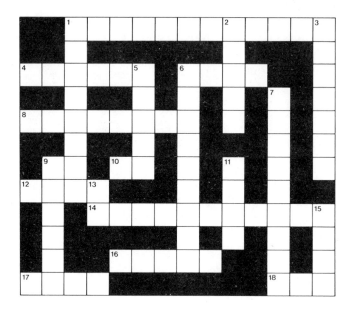

Crossword on Chapter 8

Across

1 This acid, with 12 across, gives 1 down (12)
4 A metal which floats on water (6)
6 The name given in years past to the ash formed when metals burn (4)
8 Someone who helps those who forget their lines (8)
9 North-east (1, 1)
10 Symbol for aluminium (2)
12 This metal reacts with steam to form a yellow solid (4)
14 They grow up in caves and are made of 3 down, 6 down (11)
16 See 15 down
17 A name for calcium stearate (4)
18 The water here contains minerals beneficial to health (3)

Down

1 A name meaning *water-maker* (8)
2 Colour of potassium flame (5)
3 This metal reacts with cold water (7)
5 The type of element which can displace hydrogen from water or from most acids (5)
6 Acids give carbon dioxide if added to this (9)
7 These water-softeners exchange 3 down for 4 across (9)
9 This acid gives a brown gas with 12 across or with 5 down (6)
11, 16 across   Used to test for carbon dioxide (9)
13 Symbol for caesium (2)
15, 16 across   A solution of carbon dioxide (4, 5)

9. Salts

When hydrogen atoms in an acid are replaced by metal atoms, the substance no longer behaves as an acid: it is a completely new substance with new properties, and it is called a salt.

9.1 Preparation of salts

Methods which can be used for the preparation of salts are:

(1) Direct synthesis
(2) Reaction of a metal with a dilute acid
(3) Reaction of a base with a dilute acid ⎤
(4) Reaction of a metal carbonate with a dilute acid ⎟ Methods for soluble salts
(5) Reaction of an alkali with a dilute acid ⎦
(6) Precipitation of an insoluble salt

Direct synthesis

In this method, two elements are mixed, and ground together or heated together to allow them to combine to form the required salt. It is a method which can be used for very reactive elements. In Chapter 4, iron(II) sulphide, aluminium iodide and mercury iodide were prepared by direct synthesis. You may want to revise this method by asking your teacher to demonstrate Experiment 9.1, in which zinc and sulphur combine to form zinc sulphide.

Demonstration Experiment 9.1

To combine zinc and sulphur

1. Take 2 g zinc dust and 1 g powdered sulphur on a piece of paper, and roll them together. *Do not grind*. Tip the mixture into an evaporating basin.

2. Wearing safety glasses, plunge a lighted taper into the mixture.

When you see what happens, you will have little doubt that a chemical reaction is taking place. Why do you think that, when this experiment was first done, people called the substance formed *philosopher's wool*? In fact, zinc and sulphur combine to form zinc sulphide:

$$Zn(s) + S(s) \rightarrow ZnS(s)$$

Reaction between a metal and an acid

This method is used for the preparation of soluble salts. Many metals react with dilute acids to give hydrogen and a salt of the metal and acid. For example, in Experiment 9.2:

$$\text{Zinc} + \text{Sulphuric acid} \rightarrow \text{Zinc sulphate} + \text{Hydrogen}$$

$$Zn(s) + H_2SO_4(aq) \rightarrow ZnSO_4(aq) + H_2(g)$$

This method can be used for metals which are sufficiently reactive to displace hydrogen from a dilute acid, metals above hydrogen in the activity series. Copper and lead are below hydrogen in the activity series in Table 8.2 (p. 152) and will not displace it from acids. They will, however, react with dilute nitric acid, because it is an oxidising agent as well as an acid, to give nitrates. Sodium and potassium are the most reactive metals. They are so far above hydrogen in the activity series that their reactions with acids are too vigorous to be safe. Although reaction with acids cannot be used for the most reactive and the least reactive metals, this method is useful for making many soluble salts.

Reaction between a base and a dilute acid

This method can be used for a large number of soluble salts. All metal oxides are bases and react with acids to give salts. This method can be used with the oxides of metals such as zinc, which react with dilute acids. It can also be used with the oxides of metals, such as copper and lead, which are below hydrogen in the activity series and do not react with dilute acids. For copper and lead, Method (2) can be used only with dilute nitric acid to give nitrates, and Method (3) is therefore valuable for the other salts of these metals.

Your instructions in Experiment 9.3 are for making copper sulphate:

$$\text{Copper oxide} + \text{Sulphuric acid} \rightarrow \text{Copper sulphate} + \text{Water}$$

$$CuO(s) + H_2SO_4(aq) \rightarrow CuSO_4(aq) + H_2O(l)$$

By using different metal oxides and different acids, you can use the same method to make a variety of salts. Dilute nitric acid will give nitrates, and dilute hydrochloric acid will give chlorides. Lead oxide will give lead salts, and zinc oxide will give zinc salts and so on.

Experiment 9.2

To prepare zinc sulphate

1. Take 50 cm³ dilute sulphuric acid in a Pyrex beaker. Warm. Add two pieces of granulated zinc.

2. If all the zinc reacts completely and disappears, add more. When the stream of bubbles of hydrogen stops, all the acid has been used up. There will be some zinc left over. You must use an excess of zinc, more than enough to react with all the acid, because it is important to have no acid left at the end of the reaction. Figure 9.1 (a) shows the apparatus.

3. Figure 9.1 (b) shows the next step. Filter to remove excess zinc.

4. Evaporate the filtrate until crystals of zinc sulphate just begin to form. Leave to stand, and filter off the crystals of zinc sulphate.

$$\text{Zinc} + \text{Sulphuric acid} \rightarrow \text{Zinc sulphate} + \text{Hydrogen}$$

$$\text{Zn(s)} + \text{H}_2\text{SO}_4\text{(aq)} \rightarrow \text{ZnSO}_4\text{(aq)} + \text{H}_2\text{(g)}$$

If you have not made sure that all the dilute sulphuric acid was used up in the reaction, when you start to evaporate the solution of zinc sulphate, the acid will become more concentrated, and there is a possibility of concentrated acid splashing out.

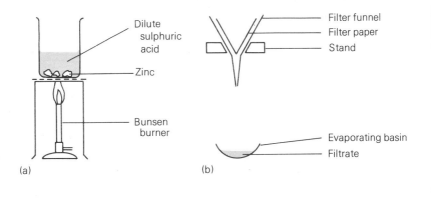

Figure 9.1   Preparation of zinc sulphate

To prepare copper sulphate crystals

1. Use the apparatus shown in Figure 9.1 (a). Into the Pyrex beaker, put 50 cm³ of dilute sulphuric acid. Warm. Add a spatula measure of copper oxide, and stir with a glass rod.

2. If all the copper oxide you have added reacts, add some more. You must have an excess of copper oxide to make sure that all the acid is used up. No gas is evolved in this reaction. You can judge when the reaction is complete because a piece of litmus paper dipped into the solution will no longer turn red when all the acid has been used up.

3. Filter to remove the excess of copper oxide. Put the filtrate of copper sulphate solution into an evaporating dish. Heat until, when you dip in a glass rod and hold it up to cool, small crystals form on the glass rod. Leave to stand so that crystallisation will take place slowly and the crystals will be large.

4. If you obtain some nicely shaped crystals which you want to keep, wrap them up in sellotape, and stick them into your book or on to a card.

Copper oxide + Sulphuric acid → Copper sulphate + Water

$$CuO(s) + H_2SO_4(aq) → CuSO_4(aq) + H_2O(l)$$

Reaction between a metal carbonate and an acid

This method can be used for the preparation of soluble salts. All metal carbonates react with dilute acids to give carbon dioxide and a solution of the metal salt, provided the salt is soluble. Heating is unnecessary. You can follow the preparation of lead nitrate described in Experiment 9.4 with other salts made from different metal carbonates and different acids:

Lead carbonate + Nitric acid → Lead nitrate + Carbon dioxide + Water

$$PbCO_3(s) + 2HNO_3(aq) → Pb(NO_3)_2(aq) + CO_2(g) + H_2O(l)$$

This method can only be used for soluble salts. If you try reacting marble chips (calcium carbonate) with dilute sulphuric acid, the reaction very quickly comes to a halt. Calcium sulphate is insoluble, and the insoluble film of calcium sulphate forming round the marble chip prevents more sulphuric acid from reaching the marble and reacting.

Reaction between an alkali and a dilute acid

The neutralisation of an acid by an alkali can be used to prepare soluble salts. Alkalis are soluble bases. An alkali reacts with an acid to form a salt and water. For example:

| Sodium | + Hydrochloric | → Sodium + Water |
| hydroxide | acid | chloride |

$$NaOH(aq) + HCl(aq) \rightarrow NaCl(aq) + H_2O(l)$$

The common alkalies are: sodium hydroxide, potassium hydroxide, calcium hydroxide and ammonia solution. Sodium and potassium are too reactive for the direct reaction with acids (Method 2) to be safely carried out, and the neutralisation method is especially valuable for the preparation of their salts. Experiment 9.5 gives a method for preparing an ammonium salt, and Experiment 9.6 gives a method of preparing a sodium salt.

Precipitation

The precipitation method is used for insoluble salts. Suppose you want to make an insoluble salt AB. You need to make a solution of a soluble salt of the metal A and a solution of a soluble salt of the acid group B. Then mix the solutions. The combined solution will contain ions of the metal A and ions of the acid group B. As soon as A ions and B ions meet, the insoluble salt AB is thrown out of solution. To help in choosing soluble salts, remember that all sodium salts are soluble, and all nitrates are soluble. The insoluble salt barium sulphate can be made by adding solutions of barium nitrate and sodium sulphate. The precipitate is filtered, washed and dried:

Barium	+	Sodium	→	Barium	+	Sodium
nitrate		sulphate		sulphate		nitrate
(solution)		(solution)		(precipitate)		(solution)

$$Ba(NO_3)_2(aq) + Na_2SO_4(aq) \rightarrow BaSO_4(s) + 2NaNO_3(aq)$$

The precipitation method is used for the preparation of various pigments. Experiment 9.7 tells you how to prepare some pigments and mix them with what the industry calls a 'vehicle'. This is a mixture of solvent and binder which carries the pigment.

To prepare lead nitrate crystals

1. Use the apparatus shown in Figure 9.2. You do not need to heat. Put 50 cm³ dilute nitric acid into a beaker. Add a spatula measure of lead carbonate. Effervescence occurs as carbon dioxide is evolved.

2. If all the lead carbonate reacts, add some more: it is important that no nitric acid is left.

3. Filter to remove the excess of lead carbonate, and evaporate the filtrate until, when you dip a glass rod into it and hold up the rod to cool, small crystals appear on the rod. Leave to stand so that crystals of lead nitrate will form slowly.

4. Filter the crystals and dry them with filter paper.

$$\begin{array}{ccccccc} \text{Lead} & + & \text{Nitric} & \rightarrow & \text{Lead} & + & \text{Carbon} & + & \text{Water} \\ \text{carbonate} & & \text{acid} & & \text{nitrate} & & \text{dioxide} \end{array}$$

$$PbCO_3(s) + 2HNO_3(aq) \rightarrow Pb(NO_3)_2(aq) + CO_2(g) + H_2O(l)$$

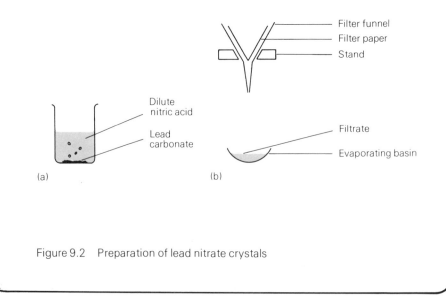

Figure 9.2   Preparation of lead nitrate crystals

To prepare ammonium sulphate

1. Take 50 cm³ dilute sulphuric acid in a beaker, as shown in Figure 9.3 (a).

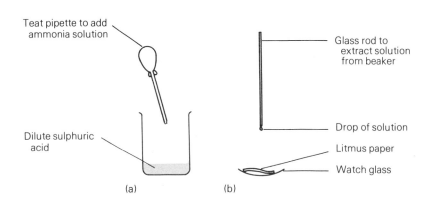

Figure 9.3   Preparation of ammonium sulphate

2. With a teat pipette, add dilute ammonia solution. Stir with a glass rod, and then take out a drop of solution and drop it onto a piece of litmus paper, as shown in Figure 9.3 (b). The paper will turn red, showing that the solution is still acid.

3. Add some more dilute ammonia solution, stir, take out a drop of solution and test it again. Continue adding more ammonia solution until the litmus turns blue, showing that the solution is alkaline. All the acid has been neutralised, and there is an excess of ammonia present. This does not matter as it will be driven off as ammonia gas and steam when you begin to heat the solution.

4. Put the solution of ammonium sulphate into an evaporating basin and heat. When the solution appears ready to crystallise, leave it to stand. Collect the crystals of ammonium sulphate formed.

$$Ammonia + Sulphuric\ acid \rightarrow Ammonium\ sulphate$$

$$2NH_3(aq) + H_2SO_4(aq) \rightarrow (NH_4)_2SO_4(aq)$$

To prepare sodium nitrate crystals

Method A You can prepare this salt by the method described in Experiment 3.4, using nitric acid instead of hydrochloric acid.

Method B This method requires the use of a burette. This is a wide glass tube with graduation marks along its length and a tap at the bottom.

1. Set up the burette as shown in Figure 9.4. It is held with a clamp and stand. Close the tap.

2. Stand the burette on the floor while you fill it, through a funnel, with dilute nitric acid. Wear safety glasses. Open the tap, so that the jet fills with acid; then close the tap.

3. Read the burette. The surface of the liquid is curved and is called a *meniscus*. You have to read the graduation mark on the burette at the bottom of the meniscus. Your eyes must be level with the meniscus. Write down the reading, V_1 cm^3 = _____.

4. Measure 25 cm^3 of dilute sodium hydroxide solution in a measuring cylinder. Use a teat pipette if necessary to adjust the volume to exactly 25 cm^3.

5. Tip the alkali into a conical flask. Add three drops of screened methyl orange indicator.

6. Stand the conical flask on a white tile. Add dilute nitric acid carefully, from the burette. Swirl the conical flask. The indicator will change from yellow to green and, finally, will suddenly turn grey at the neutral point. If the indicator turns pink, you have overshot the neutral end-point, and must repeat the titration. *Titration* is the word we use for adding liquid in a measured way.

7. Read the burette. Record the volume, V_2 cm^3 = _____.

8. Work out the volume of nitric acid needed to neutralise 25 cm^3 of sodium hydroxide solution which is $(V_2 - V_1)$ cm^3. $(V_2 - V_1)$ cm^3 = _____.

9. Wash out the conical flask. Again, measure carefully 25 cm^3 of sodium hydroxide solution in a measuring cylinder, and tip it into the flask. *Add no indicator.*

10. Refill the burette. Run in $(V_2 - V_1)$ cm^3 of nitric acid. You know this is the volume of acid needed to neutralise 25 cm^3 of the alkali.

11. Put the solution in the conical flask into an evaporating basin, and evaporate until crystals form. This product is sodium nitrate.

Sodium hydroxide + Nitric acid → Sodium nitrate + Water

$$NaOH(aq) + HNO_3(aq) \rightarrow NaNO_3(aq) + H_2O(l)$$

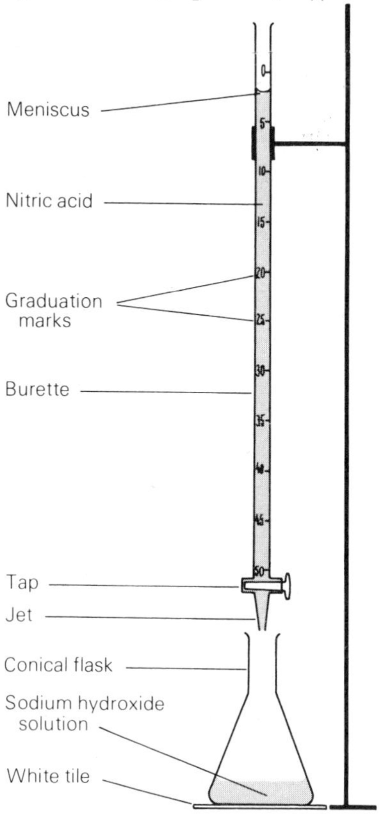

Figure 9.4   Preparation of sodium nitrate

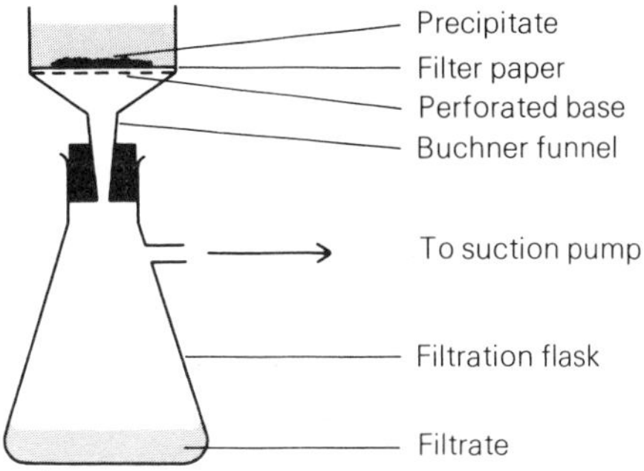

Figure 9.5   Filtration of precipitates

187

Experiment 9.7

To prepare some pigments and make them into coloured paints

1. With a measuring cylinder, measure into separate beakers the stated volumes of the two solutions required for the pigment.

2. Mix the solutions by stirring.

3. Fit a clean Buchner funnel into a filter flask, as shown in Figure 9.5, and attach a filter pump to the flask. Lay a filter paper in the funnel, wet it, and apply gentle suction. Pour the contents of the beaker into the funnel, increasing suction if necessary to draw all the liquid through. Wash the precipitate with water. (If you do not have a Buchner funnel, you can use a filter funnel and filter paper; it just takes longer.)

4. When all the water has passed through, spread the damp pigment on a paper towel to dry.

5. The volumes of solutions (all 10% concentration) to mix are:

    (a) For white pigment, basic lead carbonate, add 75 cm³ lead nitrate solution to 15 cm³ sodium carbonate solution.

    (b) For yellow pigment, lead chromate, add 15 cm³ potassium chromate solution to 75 cm³ lead nitrate solution.

    (c) For Prussian blue, Iron(III) hexacyanoferrate(II), add 20 cm³ iron(III) chloride solution to 75 cm³ potassium hexacyano-ferrate(II) solution.

    (d) For green pigment, basic copper carbonate, add 60 cm³ copper sulphate solution to 30 cm³ sodium hydrogencarbonate solution.

6. Take some of the white pigment; break it up with a spatula on a watch glass. In a test tube, add one part of glue to three parts of water. Add a few drops of this paint 'vehicle' to the pigment and mix with the spatula. Add a few more drops and mix well. Repeat until you have a paint of a nice consistency.

    Repeat the mixing with the other pigments. Try mixing a coloured pigment with white to make a pastel shade. When you have assembled a few coloured paints, take a paint brush and paint a picture with them.

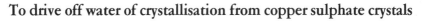

To drive off water of crystallisation from copper sulphate crystals

1. Take some copper sulphate crystals and put them into the side-arm boiling tube shown in Figure 9.6.

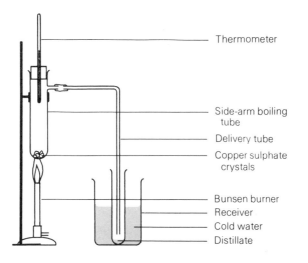

Figure 9.6   Heating copper sulphate crystals

2. Warm with a small flame. Watch the thermometer. If you heat strongly, and the temperature soars suddenly, the sudden expansion of mercury may break the thermometer.

3. Collect any distillate in the receiver.

4. What is the thermometer reading as the liquid distils over? Which liquid has this boiling point?

   What colour were the copper sulphate crystals before heating? What colour are they now?

   What shape were the copper sulphate crystals before heating? What shape are they now?

   See if you can write an equation for the reaction which has occurred.

9.2 Water of crystallisation

Many salts crystallise as hydrates, crystals which contain water. There is always a definite number of water molecules to each molecule of salt. In magnesium sulphate crystals, there are always seven molecules of water for every magnesium sulphate molecule, and the hydrate therefore has the formula $MgSO_4.7H_2O$. It is called magnesium sulphate-7-water, but is better known as Epsom salts. Copper sulphate crystals have the formula $CuSO_4.5H_2O$, and are named copper sulphate-5-water. Experiment 9.8 shows what happens when you remove water from copper sulphate-5-water. This method can be used for other hydrates, such as cobalt chloride-6-water, $CoCl_2.6H_2O$, and sodium carbonate-10-water, $Na_2CO_3.10H_2O$.

The combined water gives both colour and shape to the crystals, and it is called water of crystallisation.

9.3 Growing large crystals

The more slowly crystals form, the larger they become. One way of obtaining large crystals is to lag the beaker holding the solution to slow down the rate of cooling. Another method is to cover the vessel holding the solution with a piece of paper pierced with a few holes, to slow down the rate of evaporation. A third method is to give the solute a centre to crystallise around by adding a small crystal of solute. As solvent evaporates, and solute comes out of solution, it will crystallise round the existing crystal rather than form new crystals. Any specks of dust will also form centres for crystallisation. The solution must be filtered to get rid of these before crystallisation starts. Two substances which readily form large crystals are copper sulphate and alum.

Experiment 9.9 tells you how to grow crystals of alum. You can use this method for many other salts.

To grow a large crystal of alum

1. Alum is the common name for potassium aluminium sulphate. This forms crystals with water of crystallisation, producing $KAl(SO_4)_2.12H_2O$.

2. Prepare a solution of alum. Take a $250 \ cm^3$ beaker and put into it $150 \ cm^3$ water. Warm, but do not boil. Add alum and stir. When no more alum will dissolve, allow the solution to cool. You now have a cold, saturated solution.

3. Filter the solution into a clean beaker.

4. Take a small, well shaped crystal of alum. Suspend it by a thread from a glass rod so that the crystal hangs in the middle of the solution (see Figure 9.7).

5. Cover with a sheet of paper pierced with a few holes. Set aside in a place where the temperature is constant.

6. The solution will slowly evaporate, and the small crystal will grow in size. If small crystals form elsewhere, remove them. As the level of the saturated solution falls, top it up with more. If the solution is not saturated, the crystal will dissolve.

7. There are various ways of preserving the crystal you have grown, to prevent it losing its water of crystallisation. After drying it, you can put the crystal into a specimen bottle with a plastic top. You can coat the crystal with perspex cement or nail varnish and mount it on a card, or you can seal it into a glass tube.

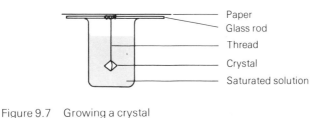

Figure 9.7   Growing a crystal

9.4 Some useful salts

Common salt

The salt we call common salt is sodium chloride. It is used to flavour foods, and it is used to preserve foods. In the soap-making industry, salt is used in the 'salting-out' process. The Solvay process for the manufacture of washing soda and baking soda uses enormous quantities of brine, and the industry is therefore located in Cheshire, where there are large salt deposits. Salt is used in the manufacture of hydrochloric acid. When brine is electrolysed, sodium hydroxide, chlorine and hydrogen are obtained. Under different conditions, sodium chlorate(I), $NaClO$, is obtained; this is the chemical in domestic bleaches. At a higher temperature, sodium chlorate(V), $NaClO_3$, is the product; this is a powerful weed killer. When solid sodium chloride is melted and electrolysed, sodium and chlorine are formed.

Silver bromide

When solutions of potassium bromide and silver nitrate are added, a pale yellow precipitate of silver bromide is formed. If the precipitate is allowed to stand, it gradually becomes a brownish colour and eventually turns black. The process takes place faster in bright sunlight. It is caused by the action of light on silver bromide, splitting up the compound to silver and bromine. Silver is formed as minute black particles.

This reaction is used in photography. A photographic film is made by adding a solution of potassium bromide and gelatin to a solution of silver nitrate. The particles of silver bromide formed are suspended in gelatin: this keeps the particles small and keeps them separate. The mixture is used to coat celluloid and allowed to set.

When the film is exposed in a camera, light falls on the film for a very short time (say a thirtieth of a second). During the exposure a very small amount of silver is formed by the action of light on silver bromide. Most silver is formed in areas where most light has fallen. When the film is developed, the solution of developer continues the process of changing silver bromide to silver started by the light. It does not change silver bromide which has not been affected by light. Developer is washed off, and then a solution of sodium thiosulphate (called *hypo*) is used to dissolve the remaining silver bromide. The film is washed and dried and is then called a *negative*. A negative is black where light fell on it because of the deposit of silver and transparent where no light fell because silver bromide has been removed. A positive must now be obtained from this negative, by the process called *printing*. The negative is placed on white paper coated with a mixture of silver bromide and gelatine. It is exposed to light, and light passes through the negative on to the light-sensitive paper. Very little light can pass through the black sections, but light can pass through the transparent sections of the negative. It gives a positive in which the dark and light patches of the negative are reversed. The paper is then developed in the same way as the film to give a print.

Plaster of Paris

Calcium sulphate is mined as gypsum, $CaSO_4.2H_2O$. When it is heated to 110 °C, it forms a compound of formula $CaSO_4.\frac{1}{2}H_2O$, which is called *plaster of Paris*. When mixed with water it sets to form a hard mass of the hydrate, $CaSO_4.2H_2O$, and expands slightly. Because of this property, it is used for making plaster casts for broken limbs.

Calcium chloride

Calcium chloride is made in large quantities as it is the waste product of the Solvay process for making sodium carbonate. It is a drying agent, and gases can be passed through a tower containing calcium chloride to dry them (except for ammonia, which combines with calcium chloride). A few lumps of calcium chloride dropped into a liquid such as benzene or toluene will remove small amounts of water from the liquid.

Washing soda and baking soda

Sodium carbonate crystallises from solution as a hydrate, $Na_2CO_3.10H_2O$. These transparent crystals are commonly called washing soda. Washing soda is used as a cleansing agent. Bath salts contain these crystals together with some colouring and perfume. You can make some in Experiment 9.10.

To make some bath salts

1. Take a beaker (250 cm³) containing 100 cm³ water. Warm it on a tripod and gauze.

2. Add sodium carbonate, stirring with a glass rod, until no more will dissolve.

3. Filter. Add a little colouring matter (for example cochineal). Leave the solution in a crystallising dish covered with a piece of paper pierced with some holes.

4. After a few days, you will have a batch of crystals. Filter, and spread them on a paper towel to dry.

5. Keep some of the crystals for Experiment 9.11.

6. Put the rest of the crystals into a bottle. Add a little perfume, and shake very gently to distribute the perfume without smashing the crystals.

7. Stopper the bottle of bath salts.

To study the action of air on washing soda crystals

1. Take some crystals of sodium carbonate hydrate, $Na_2CO_3.10H_2O$, and spread them on a petri dish.

2. Look at the crystals after an hour, after a day and again after a week.

3. Explain what has happened to the washing soda crystals. Write an equation if you can.

The reason for doing these two experiments is to find out whether there are chemical differences between sodium carbonate, Na_2CO_3, and sodium hydrogencarbonate, $NaHCO_3$.

To find the pH of solutions of sodium carbonate and sodium hydrogencarbonate

1. Put a spatula measure of sodium carbonate into a test tube three quarters full of water, and dissolve it.

2. Add two drops of universal indicator. Note the colour, and compare it with the colour chart for the indicator. Each colour has a number called the pH number. Note the pH number.

3. Repeat with sodium hydrogencarbonate.

To study the action of heat on sodium carbonate and sodium hydrogencarbonate

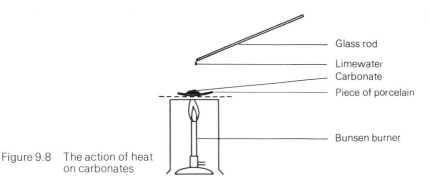

Glass rod
Limewater
Carbonate
Piece of porcelain

Bunsen burner

Figure 9.8   The action of heat on carbonates

1. Put a spatula measure of sodium hydrogencarbonate on to a piece of porcelain (or ceramic paper) as shown in Figure 9.8.

2. Warm gently, and test for carbon dioxide by holding a drop of limewater on the end of a glass rod above the heated carbonate.

3. Heat more strongly, and test again.

4. Repeat the experiment, using sodium carbonate.

5. Did either or both of the carbonates give carbon dioxide (a) on gentle heating, (b) on strong heating?

You will have found out, in Experiment 9.11, that, on standing, crystals of sodium carbonate-10-water, $Na_2CO_3.10H_2O$ (washing soda), lose water of crystallisation. This process is called *efflorescence*. The crystals eventually become sodium carbonate-1-water, $Na_2CO_3.H_2O$, which is a powdery white solid. The one remaining molecule of water of crystallisation can be driven off by heating.

Sodium hydrogencarbonate has the formula $NaHCO_3$. Carbonic acid is the weak acid formed when carbon dioxide dissolves in water. It has the formula H_2CO_3. In sodium carbonate, both hydrogen atoms have been replaced by sodium to give Na_2CO_3, whereas in sodium hydrogencarbonate only one of the hydrogen atoms has been replaced by sodium to give $NaHCO_3$. You will have found, by doing Experiments 9.12 and 9.13, that the difference between Na_2CO_3 and $NaHCO_3$ makes a big difference to the chemical behaviour of these compounds.

Sodium carbonate solution is strongly alkaline, with a pH of 11. Sodium hydrogencarbonate solution is weakly alkaline, with a pH of 9. The strongly alkaline nature of sodium carbonate solutions makes them useful. They help to remove grease and dirt from fabrics, and this is why sodium carbonate is called washing soda. Washing powders contain sodium carbonate and a whitening agent. Bath salts contain sodium carbonate and perfume and colouring material.

Sodium hydrogencarbonate is a weak alkali when in solution. It is used in indigestion powders to neutralise stomach acidity because it is a weak enough alkali to be safe to swallow. It is referred to as *sodium bicarbonate* or 'bicarb. of soda'. The difference in the pH of the solutions thus leads to very different uses for these two carbonates, sodium carbonate for washing powders and sodium hydrogencarbonate for indigestion.

You will have found that sodium hydrogencarbonate decomposes readily on gentle heating to give carbon dioxide, but sodium carbonate does not decompose on heating. This difference is responsible for the use of sodium hydrogencarbonate as a rising agent, producing carbon dioxide to make bread and cakes rise. Baking powder contains sodium hydrogencarbonate and tartaric acid, a weak acid. Sodium hydrogencarbonate is called baking soda.

You may wonder why you had to go to the trouble of heating on a piece of porcelain instead of simply heating the two solids in ignition tubes. The snag is that if you do the heating in glass, a component of glass, silicon oxide (called silica) splits up sodium carbonate to form sodium silicate. You are studying the reaction of these two carbonates with silica, not the reaction on heating, and under these conditions both compounds give carbon dioxide.

The equation for thermal decomposition of sodium hydrogen-carbonate is:

Sodium hydrogencarbonate $\rightarrow$ Sodium + Carbon + Steam
carbonate dioxide

$$2NaHCO_3(s) \rightarrow Na_2CO_3(s) + CO_2(g) + H_2O(g)$$

Questions on Chapter 9

1. Decide whether each of the following is an acid or a base or a salt:

   (a) zinc oxide   (b) limewater

   (c) ammonia   (d) carbon dioxide solution

   (e) copper sulphate   (f) lead nitrate

   (g) sulphur dioxide solution   (h) magnesium oxide.

2. What do all acids contain? What is a salt? What do we call the salts that are made from (a) hydrochloric acid, (b) nitric acid and (c) sulphuric acid?

3. Name three substances which react with sulphuric acid to give zinc sulphate. Write equations for the three reactions.

4. What is a salt?
   Describe how you would prepare:
   (a) a salt of zinc, starting from the metal,
   (b) a salt of sodium, starting from the hydroxide.

5. What do you understand by water of crystallisation? Give the names and formulae of two salts which contain water of crystallisation.

6. What is the chemical name of common salt?

   Which acid and which alkali would react to form common salt?

   Write a word equation and a chemical equation for the reaction.

   Describe in detail how you would make crystals of common salt from the acid and alkali you have chosen.

7. Write down the correct words to fill the gaps:

   Acid + Alkali   → Salt + _____
   Acid + Base   → Salt + _____
   Acid + Metal   → Salt + _____
   Acid + Carbonate → Salt + _____ + _____.

8. Describe how you would make a soluble salt other than common salt. Name the salt, name the chemicals you would start with, and describe what you would do with them to obtain crystals of the salt.

9. Lead sulphate is a salt which is insoluble in water. Name two salts which you could use to make it, and say what you would do with them.

10. Blue copper sulphate crystals are heated.

    (a) What is the final colour after heating?

    (b) What liquid can be collected from the crystals? How can you identify it?

    (c) Where does that liquid come from?

    (d) What is the name of copper sulphate without the liquid?

    (e) What colour change happens when the liquid is added to the solid left behind after heating?

11. (a) Name a black solid which reacts with warm, dilute sulphuric acid to give a blue solution.

    (b) What compound is present in plaster of Paris?

    (c) What is the name of the drying agent that is formed in the manufacture of sodium carbonate?

    (d) What is the chemical formula for bath salts?

    (e) What is sodium hydrogencarbonate used for?

    (f) How can you tell the difference between sodium carbonate and sodium hydrogencarbonate?

    (g) What light-sensitive salt is used in photographic films?

Trace this grid on to a piece of paper, and then fill in the answers.

Crossword on Chapter 9

Across

1 See 17 down
5, 14 down One starting material in the preparation of a salt (5, 5)
8 The highest point (3)
10 Roman house (5)
12, 3 down, 13 down They turn from blue to white on heating (6, 8, 8)
14 '_____ aye!' says the Scotsman (3)
16 See 4 down
17 Symbol for beryllium (2)
21 One way of making salts (14)
23 See 4 down
24 Some of these salts are fertilisers (8)

Down

2 Ending to a morning's work (5)
3 See 12 across and 11 down
4, 23 across, 16 across Lost when 12 across, 3 down, 13 down is heated (5, 2, 15)
6 A starting material in the preparation of salts (4)
7, 20 down A name for sodium carbonate-10-water (7, 4)
9 Alternatively (2)
11, 3 down Plaster of Paris (7, 8)
13 See 12 across
14 See 5 across
15, Colour of anhydrous cobalt chloride (4)
17, 1 across A use for sodium carbonate crystals (4, 5)
18 Salts must be heated very strongly before they do this (4)
19 A laboratory should be this (4)
20 See 7 down
22 Decay (3)

Index